Yingjie Liang, Wen Chen, and Wei Cai
Hausdorff Calculus

Fractional Calculus in Applied Sciences and Engineering

Volume 6

Yingjie Liang, Wen Chen, and Wei Cai

Hausdorff Calculus

Applications to Fractal Systems

DE GRUYTER

Mathematics Subject Classification 2010
Primary: 28A80, 60G22, 34A08; Secondary: 11K55, 34K37

Authors

Asst. Prof. Yingjie Liang
Hohai University
College of Mechanics and Materials
No. 8 Focheng West Road
211100 Nanjing
P.R. China
liangyj@hhu.edu.cn

Prof. Wen Chen
Hohai University
College of Mechanics and Materials
No. 8 Focheng West Road
211100 Nanjing
P.R. China
chenwen@hhu.edu.cn

Asst. Prof. Wei Cai
Hohai University
College of Mechanical and Electrical Engineering
No. 200 North Jinling Road
213022 Changzhou
P.R. China
cwprince@126.com

ISBN 978-3-11-060692-8
e-ISBN (PDF) 978-3-11-060852-6
e-ISBN (EPUB) 978-3-11-060705-5
ISSN 2509-7210

Library of Congress Control Number: 2018967902

Bibliographic information published by the Deutsche Nationalbibliothek
The Deutsche Nationalbibliothek lists this publication in the Deutsche Nationalbibliografie; detailed bibliographic data are available on the Internet at http://dnb.dnb.de.

Cover image: naddi/iStock/thinkstock
Typesetting: VTeX UAB, Lithuania
Printing and binding: CPI books GmbH, Leck

www.degruyter.com

Preface

Unlike the fractional derivative, the Hausdorff derivative, one kind of fractal derivatives (also called the local fractional derivative), introduced by Chen in 2006, is a local derivative instead of a global operation. Thus, the computing costs of the Hausdorff derivative are far less than those of the global fractional derivative, while performing almost equally well in modeling a variety of complex problems. In particular, the Hausdorff derivative diffusion equation characterizes the stretched Gaussian process in space and fractal exponential decay, also known as stretched relaxation, in time. In contrast, the fractional derivative diffusion equation underlies the Lévy statistics in space and the Mittag-Leffler function power law decay in time. The Hausdorff calculus operator in space also underlies the non-Euclidean distance and has potential applications in a variety of complex problems, and the corresponding fundamental solutions of the Hausdorff partial differential equations (PDEs) have been analytically derived and numerically verified. The methodology in this book consists of self-contained approaches and theory of calculus, statistics, geometry, metrics, and computation. The Hausdorff calculus method significantly extends the application scope of the classical calculus modeling approach under the framework of continuum mechanics to fractal materials. It is noted that nowadays there exist quite a few different definitions of fractal derivative, among which, to the best of our understanding and knowledge, the Hausdorff derivative is mathematically the simplest and numerically the easiest to implement with clear physical significance and the most real-world applications.

The Hausdoff derivative has been applied in anomalous diffusion, viscosity, creep, relaxation, ultraslow diffusion, non-Gaussian distribution, and non-Debye exponential decay. This is the first book of the Hausdorff derivative and its comprehensive applications, which are related to but evidently different from fractional calculus. This book is based on the recent research work of Chen's research group, which covers the basic theory and applications of Hausdorff fractal distance, radial basis function (RBF), statistics, stability, fundamental solutions, Hausdorff PDE modeling, and the local structural derivative. Most of the sections conclude with some additional details, intended for individual reading, and to make the book relatively self-contained. Although we did attempt to highlight connections with the existing literature, it seems inevitable that we should leave out some important contributors, for which we can only beg their forgiveness. This book is targeted at the researchers and graduate students whose interests fall in the scope of the abovementioned topics. Many interesting research issues in this area remain open. This book will guide the motivated reader to understand the essential background and current research work and to gain the insights and techniques needed to begin making their own contributions to this rapidly growing field.

https://doi.org/10.1515/9783110608526-201

The help provided by our group members and friends is gratefully acknowledged. We thank the following persons for their valuable assistance: Dr. Fajie Wang, Dr. Hong-Guang Sun, Ms. Wei Xu, Mr. Xu Yang, Mr. Xindong Hei, and Mr. Xianglong Su. Any remaining errors are the sole responsibility of the authors. Finally, the authors would like to thank their families. Your love, support, and patience make everything possible. We also acknowledge the research grants of the National Natural Science Foundation of China (Nos. 11702085, 11772121, 11702084), the Fundamental Research Funds for the Central Universities (Nos. 2019B16114, 2018B687X14), and the China Postdoctoral Science Foundation (No. 2018M630500).

Nanjing, September 2018 Yingjie Liang, Wen Chen, Wei Cai

Contents

1 Introduction

1.1 Calculus on fractals

Time and space are fundamental mathematical and physical quantities in nature. Calculus on fractals is defined based on the fractal metric [31], which is a space-time power function (x^{β}, t^{α}), where α and β are the time and space Hausdorff dimensions, respectively. The stretched relaxation [139] and the stretched Gaussian statistics [15] are considered a consequence of the fractal metric, while the classical Gaussian distribution and Brownian motion correspond to the special case of space-time fabric. The restoration of the normal diffusion implies the invariance of physical law under scale transforms and equivalence between anomalous environmental effects and scale time-space geometry, which is reminiscent of the two pillar principles of general covariance and equivalence in the general relativity [31].

Generalizing these observations, we conjecture the following two hypotheses. (1) The hypothesis of fractal invariance: the laws of physics are invariant regardless of the fractal (scale) metric space-time (coordinate systems). (2) The hypothesis of fractal equivalence: the influence of anomalous environmental fluctuations on physical behaviors equals that of the time-space transforms. The first hypothesis means that the general form of physical equations would be invariant under the fractal transformations. The second one suggests that the anomaly in physical behaviors (e. g., anomalous diffusion [135]) is caused by environmental effects (field noise) and can fully be explained and represented by the scale space-time geometry. In terms of the Hausdorff dimensions of metric space-time, the Hausdorff derivative [31] is proposed by using a space-time power function (x^{β}, t^{α}) as the kernel functions. When $\alpha = \beta = 1$, it reduces to the traditional derivative.

1.2 Hausdorff calculus versus fractional calculus

Since the 1980s, fractional calculus [75, 124] has been found to be an alternative mathematical tool to characterize complex viscoelastic behaviors [69], anomalous diffusion [135], vibration [171], system control [88], and signal processing [180], due to its unique capability in describing anomalous behavior and memory effects. However, the non-local property of fractional calculus leads to exponentially increasing costs in the numerical simulation of long-term and large-scale problems. In addition, the fractional derivative models correspond to the Mittag-Leffler decay [117] and the Lévy stable statistics [116], and they cannot well describe those dissipation processes underlying the stretched relaxation and the stretched Gaussian statistics. Unlike the fractional derivative, the Hausdorff derivative is a local differential operator instead of the global fractional derivative. Thus, its computational costs are far lower than for the global fractional derivative.

https://doi.org/10.1515/9783110608526-001

A variety of definitions of fractal derivative have been proposed in recent years. Zähel [203] presented the concept of the local fractional derivative based on the idea of fractal geometry, which is formally similar to the definition given by Jumarie [97]. Tarasov [178] also proposed a kind of fractal derivative definition. On the basis of Tarasov and Jumarie's work, Li and Ostoja-Starzewski [110] developed a definition of fractal derivative describing the fractal materials.

Independently, Chen [31] proposed the definition of the Hausdorff derivative based on the fractal metric. Recently, Balankin and Elizarraraz [8] pointed out that the Hausdorff derivative is in fact equivalent to the fractal derivative proposed by Li and Ostoja-Starzewski. It is also known that the Hausdorff derivative has a clear significance in fractal geometry [111] and can well be interpreted in practical physics processes.

Though there exist quite a few different definitions of fractal derivative, the Hausdorff derivative has a definite statistical interpretation and many real-world applications to a wide range of problems, such as anomalous diffusion [45], magnetic resonance imaging (MRI) [115], heat conduction [155], viscoelasticity [21, 23], and water transport [177].

1.3 Hausdorff fractal distance, RBF, statistics, stability, fundamental solution

1.3.1 Hausdorff fractal distance

In the Euclidean space $\mathbf{R}^n$, the distance between two points is usually named Euclidean distance. However, fractal media are considered to be discontinuous, thus the traditional distance may be not suitable. Usually, the method of scaling transformation is employed to investigate the fractal media, which is also known as the fractal metric. On the basis of scaling transformation, we consider the following mapping: $\mathbf{X}(x_1, x_2, \ldots, x_n) \to \mathbf{X}'\left(x_1^\beta, x_2^\beta, \ldots, x_n^\beta\right)$. The distance between two points with the new metric in the three-dimensional case can be easily defined [22], which is detailed in Chapter 2.

In this way, an arbitrary element in the fractal metric $\mathbf{X}'\left(x_1^\beta, x_2^\beta, \ldots, x_n^\beta\right)$ is ensured to be a real number. It coincides with the original concept of fractal metric proposed in [31]. The Hausdorff fractal distance is a generalization of the scaling transformation. It has been widely recognized that the fractal dimensionality varies in different directions, which means the value of β is not always a constant. The non-Euclidean Hausdorff fractal distance is also usable for anisotropic fractal media.

1.3.2 Radial basis function (RBF)

Based on the Hausdorff fractal distance, various numerical methods, such as the finite element methods (FEMs) [141] and the boundary element methods (BEMs) [131], can be easily used to solve physical problems governed by the Hausdorff derivative equations. Compared with the mesh-based techniques such as the FEM and the BEM, the meshless-based methods [99, 100] have certain advantages in terms of computational costs and domain complexity. The Kansa method [101], as a global meshless method, is a radial basis function collocation technique and compares favorably with the classical methods, for its superior convergence, meshless merit, and easy implementation. The method has been successfully applied to various scientific and engineering problems governed by integer- and fractional-order derivative equations, such as radioactive transport [103], electromagnetism [77], and heat conduction [74].

In Chapter 5, four commonly used radial basis functions (RBFs) [38], i. e., multiquadric (MQ), inverse multiquadric (IMQ), Gaussian (GA), and polyharmonic splines (PS), with the Hausdorff fractal distance are presented, and they are compared to find an optimal algorithm. An efficient technique to choose the optimal shape parameters of the RBFs with the Hausdorff fractal distance is proposed with the aid of the leave-one-out cross-validation algorithm [157]. We also apply the singular boundary method (SBM) [33] and the present RBF method to numerically solve the Hausdorff derivative Laplace equations and Hausdorff derivative Poisson equations.

1.3.3 Statistics

In a statistical description or phenomenological modeling, the fractal has long been considered responsible for anomalous physical behaviors and is claimed to have links with fractional derivatives, Lévy statistics, fractional Brownian motion, and empirical power law scaling.

From the perspective of statistics, a fundamental solution of the space and time Hausdorff derivative transport equations appears as a stretched Gaussian distribution and its characteristic function has a stretched exponential form. The fractal derivative diffusion equation describes fractional Brownian motion, which is a fractal curve. The corresponding Hausdorff fractal dimension of the diffusion trajectory is a linear function of the time derivative order [115]. In real applications, the stretched Gaussian signals [195] can be easily depicted by the fractal derivative relaxation equation or the stretched exponential function, which can be considered the Fourier transform of the stretched Gaussian distribution. More details are given in Chapter 3.

1.3.4 Stability

There are many kinds of non-integer order differential dynamical systems which are proved to be chaotic or hyper-chaotic by the qualitative theory of differential equations [108]. For such chaotic systems, it is not possible to control the errors of their numerical simulations caused by the errors in the initial conditions. Meanwhile, chaotic behavior also leads to instability, which is not acceptable for some engineering applications. Therefore, it is necessary to establish stability criteria for dynamical systems. As a special case of Lyapunov stability, the concept of exponential stability is important because it guarantees that non-zero solutions of the system will decay at an exponential rate [95]. However, previous studies have shown that a Lyapunov stable system does not necessarily decay exponentially [55, 192].

In recent years, continuously growing attention has been focused on investigating the stability of fractional-order systems. Furthermore, the concepts of Mittag-Leffler stability and generalized Mittag-Leffler stability were proposed for non-linear fractional-order systems [112]. It is noted that in many real-world physical processes, the Hausdorff differential dynamic systems underlie the stretched exponential decay phenomena, which are totally different from the Mittag-Leffler decay. Hence it is necessary to establish the stability criteria for the Hausdorff dynamical systems and give a qualitative evaluation of the convergence rate of such systems. Then the criteria shall be presented as a framework on how to stabilize the Hausdorff dynamical systems. More details can be found in Chapter 4.

1.3.5 Fundamental solution

Reference [31] gives the fundamental solution of the one-dimensional Hausdorff derivative transient diffusion equation via the time-space fractal metric, namely, the Hausdorff fractal distance. Chapter 2 employs this non-Euclidean metric to derive the fundamental solution of the general n-dimensional diffusion equation and to give fundamental solutions for the most commonly used Hausdorff differential operators.

2 Hausdorff diffusion equation

2.1 Velocity on fractal

Considering a particle moving in porous media with the fractal behaviors of delay and jump, the distance can be characterized in fractal time by [24, 35]

$$l(\tau) = v \cdot (\tau - t_0)^p, \tag{2.1}$$

where v is the velocity, l denotes the distance, τ and t_0 represent the current and initial time, and p is the time fractal dimensionality. Considering the variable velocity problem, the distance can be formulated as

$$l(t) = \int_{t_0}^{t} v(\tau) d(\tau - t_0)^p, \tag{2.2}$$

and the velocity on fractal media can be defined as

$$\frac{dl}{d(t - t_0)^p} = \lim_{t \to t'} \frac{l(t) - l(t')}{(t - t_0)^p - (t' - t_0)^p} = \frac{1}{p(t - t_0)^{p-1}} \frac{dl}{dt}, \tag{2.3}$$

where t and t', respectively, represent the final and internal time instances. Equation (2.3) is the limit definition of the Hausdorff derivative, and the general formulation of the Hausdorff derivative [31] is

$$\frac{dl}{dt^p} = \lim_{t \to t'} \frac{l(t) - l(t')}{t^p - t'^p} = \frac{1}{pt^{p-1}} \frac{dl}{dt}. \tag{2.4}$$

The only difference between equations (2.3) and (2.4) is that the definition in equation (2.3) considers the initial time.

According to the definition in equation (2.1), the location of the particle at time τ can be formulated as

$$l(\tau) = v(t - t_0)^p - v(t - \tau)^p, \tag{2.5}$$

where t is the final time, the first term on the right-hand side represents the total movement distance, and the second term represents the movement distance between t and τ. Similarly, for a variable velocity problem, the differential form of equation (2.5) can be shown as

$$dl = -vd(t - \tau)^p. \tag{2.6}$$

The integral formulation of equation (2.6) can be formulated as

$$l(t) = \int_{t_0}^{t} -v(\tau) d(t - \tau)^p. \tag{2.7}$$

https://doi.org/10.1515/9783110608526-002

The classical integral definition of the Riemann–Liouville fractional derivative is [24]

$$S(t) = \frac{1}{\Gamma(p)} \int_{t_0}^{t} \frac{v(\tau)}{(t-\tau)^{1-p}} d\tau = \frac{1}{p\Gamma(p)} \int_{t_0}^{t} -v(\tau) d(t-\tau)^p, \tag{2.8}$$

where $\Gamma(\cdot)$ represents the gamma function. It is noted that equation (2.8) is identical to equation (2.7) without considering the constant before the integration.

2.2 Physical interpretation of the Hausdorff derivative

It has been well known that the mass m of a one-dimensional medium can be characterized by

$$m = \int_{x_1}^{x_2} \rho(x) dx, \tag{2.9}$$

where $\rho(x)$ represents the linear density, and x_1 and x_2 ($x_2 > x_1$) are the locations of the two ends of the media. In terms of the fractal media, the mass m is directly related to the power function of the length metric [110, 178], i. e.,

$$m \propto l^{\beta}, \tag{2.10}$$

where β is the fractal dimension of the given medium. Equation (2.9) can be formulated as

$$m = \int_{x_1}^{x_2} \rho(x) dx^{\beta}. \tag{2.11}$$

The following Hausdorff derivative in space has been proposed by Chen [31]:

$$\frac{dm}{dx^{\beta}} = \lim_{x \to x'} \frac{m(x) - m(x')}{x^{\beta} - x'^{\beta}} = \frac{1}{\beta x^{\beta-1}} \frac{dm}{dx}. \tag{2.12}$$

Similarly, the time Hausdorff derivative is stated as

$$\frac{dl}{dt^{\alpha}} = \lim_{x \to x'} \frac{l(t) - l(t')}{t^{\alpha} - t'^{\alpha}} = \frac{1}{\alpha t^{\alpha-1}} \frac{dl}{dt}, \tag{2.13}$$

where α is the fractal dimensionality in time and l represents the length. It is obvious that equation (2.11) is the integral form of the space Hausdorff derivative, which in turn validates the ability of the Hausdorff derivative to characterize fractal media. Thus, the order of the space Hausdorff derivative is directly related to the fractal dimensionality. Moreover, equation (2.12) is formally similar to the definition of fractal derivative proposed by Tarasov [178] without the absolute value sign, which ensures the physical meaning of the order of the Hausdorff derivative.

The Hausdorff derivatives, i. e., equations (2.12) and (2.13), are firstly proposed by using the following metric transforms [31]:

$$\begin{cases} \hat{t} = t^{\alpha}, \\ \hat{x} = x^{\beta}. \end{cases} \tag{2.14}$$

It should be pointed out that the uniform mass of the three-dimensional fractal media is defined as [110]

$$m = \rho l^{\beta} = \rho l^{\beta_1+\beta_2+\beta_3} = \rho l^{\beta_1} l^{\beta_2} l^{\beta_3}, \tag{2.15}$$

where β_1, β_2, and β_3 represent the fractal dimensionality along three different directions.

2.3 Fundamental solution of the Hausdorff diffusion equation

2.3.1 Hausdorff fractal distance

For any set X, a function $d : X \times X \to \mathbb{R}^+$ is called a metric if it satisfies three conditions, i. e., $d(x,y) = 0$ if and only if $x = y$, $d(x,y) = d(y,x)$ for all $x, y \in X$, and $d(x,y) \le d(x,z) + d(z,y)$ for all $x, y, z \in X$. The pair (X, d) is called a metric space.

Our interest lies not in the general theory of metric spaces, but in dealing with the classic real linear space $\mathbb{R}^n$, where n denotes the spatial topological dimension. Taking $n = 3$ as an example, a point is represented by an ordered triplet ($\mathbf{x} = (x_1, x_2, x_3)$; $\mathbf{y} = (y_1, y_2, y_3)$; etc.). The general form of the metric for the two points $\mathbf{x}$ and $\mathbf{y}$ can be expressed as [7]

$$d(\mathbf{x},\mathbf{y}) := \left(\sum_{s=1}^{n} d_s^2(\mathbf{x},\mathbf{y}) \right)^{\frac{1}{2}}, \tag{2.16}$$

where $d_s(\mathbf{x},\mathbf{y}) := |\mu_s(\mathbf{x}) - \mu_s(\mathbf{y})|$ and $\mu_s(\bullet) = \varphi(\bullet_s)$, with $\varphi : \mathbb{R} \to \mathbb{R}$ being a differentiable function. One can check that this is indeed a metric on $\mathbb{R}^n$. Equation (2.16) in fact introduces a definition of a non-Euclidean metric. Nowadays, the two most popular definitions of the metric in fractal continuum are $\varphi(x_s) := \operatorname{sign}\{x_s\}|x_s|^{\gamma_s}$ [6] and $\varphi(x_s) := |1 + x_s/l_0|^{\gamma_s}$, $x_s \ge 0$ [8, 111], where γ_s denotes a scaling exponent and l_0 is the lower cutoff of fractal behavior along the Cartesian axis x_s.

Based on the two hypotheses of fractal invariance and fractal equivalence, Chen [31] introduced the concept of the fractal metric in one-dimensional topological fractal media

$$\begin{cases} \Delta\hat{t} = \Delta t^{\alpha}, \\ \Delta\hat{x} = \Delta x^{\beta}, \end{cases} \tag{2.17}$$

where α and β, respectively, denote the time and space fractal dimensionality. Obviously, the above metric is a non-Euclidean metric in a one-dimensional continuum.

In order to distinguish the above fractal metric from the fractal metric proposed by Balankin et al. [7], which is more general and includes the metric (2.17) as a special case, the metric (2.17) is called the Hausdorff fractal distance (metric). It is straightforward to obtain the Hausdorff fractal distance expression of two points **x** and **y** for the *n*-dimensional isotropic media [7, 36], i. e.,

$$\begin{cases} \Delta\hat{t} = \Delta t^{\alpha} = t^{\alpha} - t_0^{\alpha}, \\ r^{\beta} = \left(\sum\limits_{s=1}^{n} (x_s^{\beta} - y_s^{\beta})^2 \right)^{\frac{1}{2}}, \end{cases} \tag{2.18}$$

where β denotes the spatial fractal in each dimension direction of isotropic media. It can be seen that the Hausdorff fractal distance is a special case of (2.18). Furthermore, the Hausdorff fractal distance can be reduced to the classical Euclidean distance if $\beta = 1$. The metric expression (2.18) is reduced to the metric (2.17) by setting the initial instance $t_0 = 0$ and the source point at the origin $y_s = 0$ in the one-dimensional case.

It has been widely recognized that the fractal dimensionality varies in different directions, which means that the value of β is not always a constant. Without the loss of generality, the fractal distance for anisotropic fractal media should be reformulated as

$$d(\mathbf{X}', \mathbf{Y}') = r^{\beta} = \sqrt{(x_1^{\beta_1} - y_1^{\beta_1})^2 + (x_2^{\beta_2} - y_2^{\beta_2})^2 + (x_3^{\beta_3} - y_3^{\beta_3})^2}, \tag{2.19}$$

where β_1, β_2, and β_3 are the fractal orders along the three directions. It is clear that equation (2.19) is the generalized non-Euclidean Hausdorff fractal distance for anisotropic fractal media. Figure 2.1 displays the Hausdorff fractal distance between

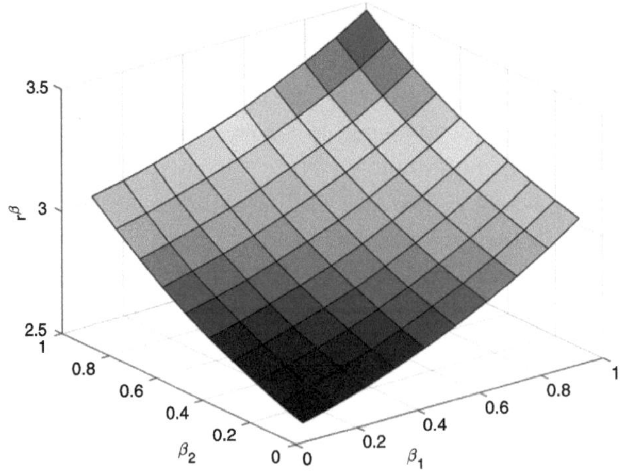

Figure 2.1: Hausdorff fractal distance between **A** (1, 1, 1) and **B** (3, 3, 3).

two test points **A** (1, 1, 1) and **B** (3, 3, 3) with varying β_1 and β_2 and where $\beta_3 = 1$ [22]. It is clear that the value of the Hausdorff fractal distance converges to the Euclidean distance with $\beta_1 \to 1$ and $\beta_2 \to 1$.

2.3.2 Fundamental solution

The Hausdorff derivative diffusion equation in one topological dimension is given by [31]

$$\frac{\partial u}{\partial t^\alpha} = D\left(\frac{\partial u}{\partial x^\beta}\left(\frac{\partial u}{\partial x^\beta}\right)\right), \tag{2.20}$$

where D represents the diffusion coefficient. By using the Hausdorff fractal distance (2.18), reference [31] gives the fundamental solution of the above diffusion equation [47], i. e.,

$$u^*(t,x) = \frac{H(t-\tau)}{(4\pi D\Delta t^\alpha)^{1/2}} \exp\left(-\frac{(\Delta x^\beta)^2}{4D\Delta t^\alpha}\right), \tag{2.21}$$

where H is the Heaviside function. It is clear that this fundamental solution underlies the stretched Gaussian distribution versus the Gaussian distribution of the normal diffusion equation.

Similarly, the fundamental solution of the n-dimensional Hausdorff derivative diffusion equation is given by [47]

$$u^*(t,x,y,z) = \frac{H(\Delta t)}{(4\pi D\Delta t^\alpha)^{n/2}} \exp\left(-\frac{(r^\beta)^2}{4D\Delta t^\alpha}\right). \tag{2.22}$$

By comparing the fundamental solution (2.21) of the Hausdorff derivative diffusion equation with the corresponding classical integer-order fundamental solution, it is easy to see that the Hausdorff fundamental solution essentially uses the non-Euclidean Hausdorff fractal metric.

2.4 Fundamental solutions of the other Hausdorff partial differential equations

Table 2.1 lists the fundamental solutions of the most often encountered Hausdorff differential operators in the topological two- and three-dimensional cases, where D is the diffusivity coefficient, λ the wave number, r^β the Hausdorff fractal distance between the source and the collocation points, $\mathbf{v}$ and $\mathbf{r}^\beta$ the velocity vector and distance vector, Δt^α the Hausdorff fractal time interval, H the Heaviside step function, and $H_0^{(2)}$ and K_0 Hankel and modified Bessel functions of order zero, respectively. The verification of these fundamental solutions is mathematically simple and is omitted here.

Table 2.1: Fundamental solutions for the most commonly used Hausdorff differential operators on the Hausdorff fractal distance of two and three topological dimensions [47].

Operator	Two-dimensional	Three-dimensional
Δ	$-\frac{1}{2\pi}\ln r^{\beta}$	$\frac{1}{4\pi r^{\beta}}$
$\Delta+\lambda^2$	$-\frac{i}{4}H_0^{(2)}(\lambda r^{\beta})$	$\frac{e^{-i\lambda r^{\beta}}}{4\pi r^{\beta}}$
$\Delta-\lambda^2$	$\frac{1}{2\pi}K_0(\lambda r^{\beta})$	$\frac{e^{j\lambda r^{\beta}}}{4\pi r^{\beta}}$
$D\Delta-\frac{\partial}{\partial t}$	$\frac{H(\Delta t)}{4\pi D\Delta t^{\alpha}}e^{-(r^{\beta})^2/4D\Delta t^{\alpha}}$	$\frac{H(\Delta t)}{(4\pi D\Delta t^{\alpha})^{3/2}}e^{-(r^{\beta})^2/4D\Delta t^{\alpha}}$
$D\Delta-\mathbf{v}\bullet\nabla-\frac{\partial}{\partial t}$	$\frac{H(\Delta t)}{4\pi D\Delta t^{\alpha}}e^{-(r^{\beta})^2/4D\Delta t^{\alpha}-\lvert\mathbf{v}\rvert^2\Delta t^{\alpha}/4D+\mathbf{v}\bullet\mathbf{r}^{\beta}/2D}$	$\frac{H(\Delta t)}{(4\pi D\Delta t^{\alpha})^{3/2}}e^{-(r^{\beta})^2/4D\Delta t^{\alpha}-\lvert\mathbf{v}\rvert^2\Delta t^{\alpha}/4D+\mathbf{v}\bullet\mathbf{r}^{\beta}/2D}$

It can be seen from Table 2.1 that the derived Hausdorff fundamental solutions are based on the non-Euclidean Hausdorff fractal metric. The fundamental solutions of the classical partial differential equations are the special limiting case when the fractal dimensionality $n\beta$ tends to topological dimensionality.

3 Statistics on fractals

3.1 Stretched Gaussian

The Hausdorff derivative diffusion equation is a mathematical tool used to describe the anomalous diffusion propagator $p(x,t)$ in complex media [31]. We have

$$\frac{\partial p(x,t)}{\partial t^{\alpha}} = D_{\alpha,\beta}\frac{\partial}{\partial x^{\beta}}\left(\frac{\partial p(x,t)}{\partial x^{\beta}}\right), \quad t > 0,\ -\infty < x < \infty, \tag{3.1}$$

where $p(x,t)$ is the probability density function of the propagator located in an infinitesimal neighborhood dx centered in x at time t, with the initial value $p(x,0) = \delta(x)$. Here, $0 < \alpha \leq 1$, $0 < \beta \leq 1$ are the respective orders of the Hausdorff derivative in time and space, and D is the generalized diffusion coefficient $(m^{2\beta}/t^{\alpha})$.

The respective definitions of Hausdorff derivative in space and time are

$$\frac{\partial p(x,t)}{\partial x^{\beta}} = \lim_{x_1 \to x}\frac{p(x_1,t) - p(x,t)}{x_1^{\beta} - x^{\beta}}, \tag{3.2}$$

$$\frac{\partial p(x,t)}{\partial t^{\alpha}} = \lim_{t_1 \to t}\frac{p(x,t_1) - p(x,t)}{t_1^{\alpha} - t^{\alpha}}. \tag{3.3}$$

We note that equation (3.1) is time- and space-dependent and can be restated as a normal diffusion equation by using the following metric transforms [31]:

$$\begin{cases} \hat{t} = t^{\alpha}, \\ \hat{x} = x^{\beta}. \end{cases} \tag{3.4}$$

The metric transforms are proposed based on the two hypotheses of fractal invariance and fractal equivalence [31]. The first hypothesis assumes the laws of physics are invariant under the fractal transformations, and the second suggests fluctuations in the statistics of the diffusion are equivalent to the fractal time-space transforms.

Then solving equation (3.1) yields a stretched Gaussian distribution [31]. We have

$$p(x,t) = \frac{1}{\sqrt{4\pi D_{\alpha,\beta}t^{\alpha}}}\exp\left(-\frac{x^{2\beta}}{4D_{\alpha,\beta}t^{\alpha}}\right). \tag{3.5}$$

The MSD is $\langle x^2(t)\rangle \sim t^{(3\alpha-\alpha\beta)/2\beta}$ [45]. In this representation, the Hausdorff derivative orders α and β can characterize the sub-diffusion and super-diffusion. In particular, when $\beta = 1$, the Hausdorff derivative diffusion equation describes the fractional Brownian motion, which is a fractal curve. The corresponding Hausdorff fractal dimension of the diffusion trajectory is $F = 2 - \alpha/2$, and it follows that $1.5 \leq F < 2$ [64]. Figures 3.1 and 3.2 show the probability density function of the stretched Gaussian distribution

https://doi.org/10.1515/9783110608526-003

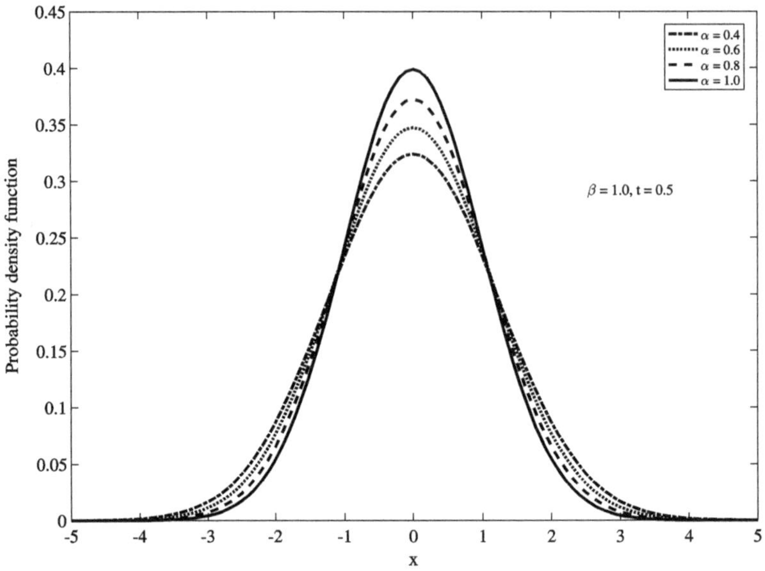

Figure 3.1: Probability density function of the stretched Gaussian distribution for the four cases of $\alpha = 0.4, 0.6, 0.8, 1.0$.

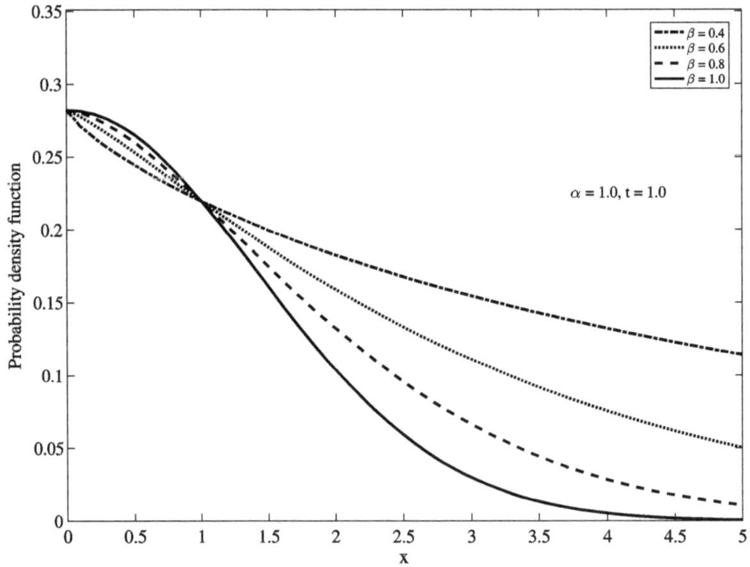

Figure 3.2: Probability density function of the stretched Gaussian distribution for the four cases of $\beta = 0.4, 0.6, 0.8, 1.0$.

for varying values of α and β, respectively. It can be observed from the two figures that the smaller the values of the orders are, the fatter the curves and the heavier the tails of the probability density function are.

In the case of $\alpha = 1$ and $\beta = 1$, equation (3.5) degenerates into the classical Gaussian distribution, i. e.,

$$p(x,t) = \frac{1}{\sqrt{4\pi D_{1,1}t}} \exp\left(-\frac{x^2}{4D_{1,1}t}\right). \tag{3.6}$$

It is noted that the basis of equation (3.6) is clear from the point of statistical mechanics, which is totally different from the Lévy stable distributions and the Mittag-Leffler distribution.

3.2 Stretched exponential decay

The characteristic function of $p(x,t)$ in equation (3.6), i. e., the Fourier transform of equation (3.6), is given by

$$p(q,t) = \exp(-D_{\alpha,\beta} q^{2\beta} t^{\alpha}), \tag{3.7}$$

which describes the relaxation for each spatial frequency (i. e., wave number) q. Euation (3.7) is a stretched exponential decay, which is also called non-Debye decay or Kohlrausch–Williams–Watts (KWW) stretched exponential relaxation. When $\alpha = 1$ and $\beta = 1$, the characteristic function becomes the classical exponential function,

$$p(q,t) = \exp(-D_{1,1} q^2 t). \tag{3.8}$$

Figures 3.3 and 3.4 give the curves of stretched exponential decay for varying values of α and β, respectively. It can be observed from the two figures that the smaller the

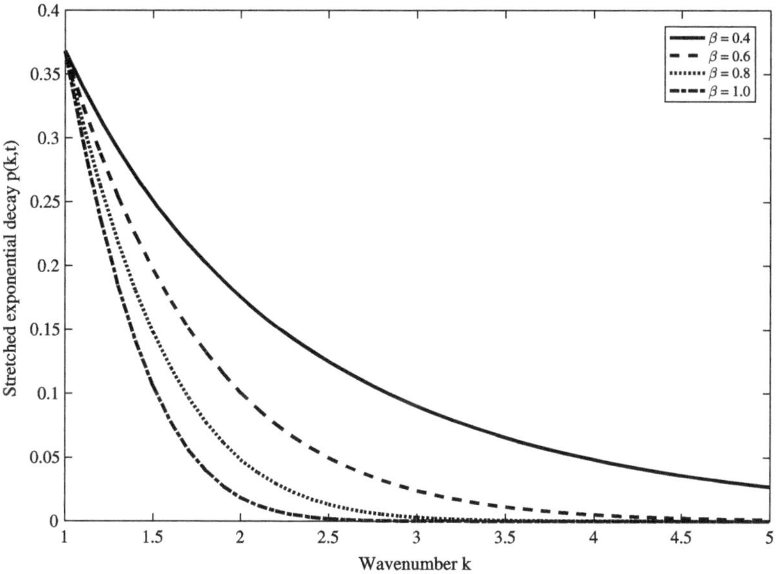

Figure 3.3: Stretched exponential decay in terms of wave number for the four cases of $\beta = 0.4, 0.6, 0.8, 1.0$.

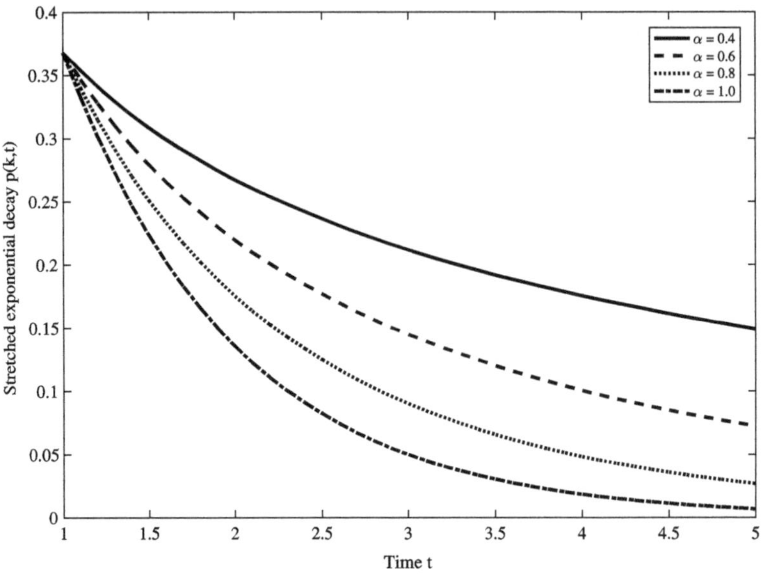

Figure 3.4: Stretched exponential decay in terms of time for the four cases of $\alpha = 0.4, 0.6, 0.8, 1.0$.

values of the derivative order is, the slower the signal decay is. The two orders can serve as markers to distinguish between different stretched exponential decays.

3.3 Feasibility study on the least square method for fitting non-Gaussian noise data

3.3.1 Non-Gaussian noise and the least square method

Non-Gaussian noise is universal in nature and engineering [67, 138, 145]. In recent decades, non-Gaussian noise has widely been studied, especially in signal detection and processing [83], theoretical model analysis [204], and error statistics [30]. It is known that the Gaussian distribution is the mathematical precondition to use the least square method [17]. However, it is often directly used to process non-Gaussian noise data, which may give wrong estimates [29].

Generally, non-Gaussian noise has detrimental influence on the stability of a system, and it also can stimulate systems to generate ordered patterns [144]. To our best knowledge, Lévy noise and stretched Gaussian noise are two kinds of typical non-Gaussian noise which are frequently used in fractional and fractal calculus [56, 154]. Lévy noise has been observed in many complex systems, such as turbulent fluid flows [169], signal processing [53], financial time series [168], and neural networks [163]. The parameter estimation for stochastic differential equations driven by Lévy noise have also been investigated [121]. Compared to Lévy noise, the stretched Gaussian noise is less studied, but its corresponding stretched Gaussian distribution has been explored [118], such as in the motion of flagellate protozoa [3], scrape-off layer (SoL) interchange turbulence simulation [176], and anomalous diffusion of particles with external force [153]. It also should be pointed out that processing of non-sinusoidal signals or sound textures has become an important research topic [104], and the derived algorithms significantly improve the perceptual quality of stretched noise signals [118].

It is well known that the least square method is a standard regression approach to approximate the solutions of over-determined systems, which is most frequently used in data fitting [106] and parameter estimation [59]. The core concept of the least square method is to identify the best match for the system by minimizing the square error. Suppose that the data points are $(x_1, y_1), (x_2, y_2), (x_3, y_3), \ldots, (x_n, y_n)$, where x represents the independent variable and y is the dependent variable. The fitting error d characterizes the distance between y and the estimated curve $f(x)$, i. e., $d_1 = y_1 - f(x_1), \ldots, d_n = y_n - f(x_n)$. The best fitting curve is obtained by minimizing the square error $\delta = d_1^2 + d_2^2 + \cdots + d_n^2 = \sum_{i=1}^{n}[y_i - f(x_i)]^2$, where the errors d_i are usually modeled by Gaussian distribution [49].

3.3.2 Generation of noise data

To examine the feasibility of the least square method in fitting non-Gaussian noise data, we generate the non-Gaussian random numbers as the noise and then add different levels of the noise to the exact values of the selected functions, including linear, polynomial, and exponential equations, to obtain the observed values. By using the least square method, the maximum absolute and mean square errors are calculated and compared in the Gaussian and non-Gaussian applications.

Field data are often polluted by noise [89], and Gaussian noise is the classical one. Its probability density function obeys Gaussian distribution. But several types of noise data obey non-Gaussian distribution [181]. For this study, the noise data are obtained based on the abovementioned Gaussian and non-Gaussian random variables. Chambers, Mallows, and Stuck proposed the CMS method in Lévy random variable simulation [28], which is the fastest and most accurate method. By using the CMS method, some variables need to be defined first. We have

$$V = \pi\left(U_1 - \frac{1}{2}\right), \quad W = -\ln U_2, \quad L = \left\{1 + \beta^2 \tan^2\left(\frac{\pi\alpha}{2}\right)\right\}^{1/2\alpha},$$
$$\theta_0 = \frac{1}{\alpha}\arctan\left(\beta \tan\left(\frac{\pi\alpha}{2}\right)\right),$$

where U_1 and U_2 are two independent uniform distributions on interval $(0, 1)$.

When $\alpha \neq 1$, the Lévy random number is

$$X = L\frac{\sin\{\alpha(V + \theta_0)\}}{\{\cos(V)\}^{1/\alpha}} \cdot \left[\frac{\cos(V - \alpha(V + \theta_0))}{W}\right]^{(1-\alpha)/\alpha}. \tag{3.9}$$

When $\alpha = 1$,

$$X = \frac{2}{\pi}\left\{\left(\frac{\pi}{2} + \beta V\right)\tan V - \beta \ln\left(\frac{\frac{\pi}{2W}\cos V}{\frac{\pi}{2} + \beta V}\right)\right\}. \tag{3.10}$$

The general Lévy random numbers can be obtained based on some known properties [191]. For the stretched Gaussian distribution, we use the acceptance rejection method to generate its random numbers [184].

3.3.3 Two classes of models

The estimation of both linear and non-linear functions is considered by using the least square method, in which the model function $f : R^m \to R$ is estimated as

$$y_i = f(x_i) + \varepsilon_i \quad (i = 1, 2, \ldots, n), \tag{3.11}$$

where n is the number of observations, $y_i \in R$ is the response variable, $x_i \in R^m$ is the explanatory variable, and the noise

$$\varepsilon_i = \text{rand} \times a\% \quad (a = 1, 5, 10, 15, 20). \tag{3.12}$$

To have a better comparison, we consider the case $n = 200$. Then the observed values $y_1, y_2, \ldots, y_{200}$ can be constructed by adding the values of the random numbers to the exact values of the selected functions, including linear, polynomial, and exponential equations. Finally the maximum absolute error and the mean square error are calculated for the above different cases by the least square method. The following abbreviations are given for convenience:

FA: Gaussian noise least square error fitting, $\mu = 5$, $\sigma = 0.5$.
FB: Lévy noise least square error fitting, $\alpha = 1.8$, $\beta = 0$, $\mu = 0$, $\sigma = 1$.
FC: Stretched Gaussian noise least square error fitting, $\beta = 2.5$, $a = 1$, $\sigma = 3$.
$F(x_i)$: The regression equation that is obtained by using the least square method.
Rerr1: Maximum absolute error: $\max|F(x_i) - f(x_i)|$.
Rerr2: Mean square error: $\sqrt{\frac{\sum_{i=1}^{n}\{F(x_i)-f(x_i)\}^2}{n}}$.

a) The simplest typical model is the linear function

$$f(x) = ax + b, \quad (a \neq 0). \tag{3.13}$$

Here we select the following case as an example:

$$f(x) = 5x. \tag{3.14}$$

Tables 3.1 to 3.5 give the estimated parameters and the errors for five different levels of noise in the linear case. We observe that the Gaussian noise fitting data maxi-

Table 3.1: The estimated results for 1 % noise in the linear case.

function	a	b	Rerr1	Rerr2
f	5	0	0	0
FA	4.9901	0.0005	0.0005	0.0003
FB	4.9973	−0.0073	0.0076	0.0075
FC	5.0087	−0.0011	0.0011	0.0007

Table 3.2: The estimated results for 5 % noise in the linear case.

function	a	b	Rerr1	Rerr2
f	5	0	0	0
FA	5.0195	−0.0009	0.0010	0.0006
FB	5.0138	0.0102	0.0116	0.0109
FC	5.0635	−0.0119	0.0109	0.0089

Table 3.3: The estimated results for 10 % noise in the linear case.

function	a	b	Rerr1	Rerr2
f	5	0	0	0
FA	5.1424	−0.0032	0.0032	0.0016
FB	4.9555	0.0104	0.0104	0.0089
FC	5.1424	−0.0103	0.0103	0.0082

Table 3.4: The estimated results for 15 % noise in the linear case.

function	a	b	Rerr1	Rerr2
f	5	0	0	0
FA	4.8624	0.0064	0.0074	0.0040
FB	4.9193	0.0482	0.0482	0.0442
FC	5.1177	−0.0322	0.0322	0.0266

Table 3.5: The estimated results for 20 % noise in the linear case.

function	a	b	Rerr1	Rerr2
f	5	0	0	0
FA	4.8533	0.0177	0.0177	0.0112
FB	5.0199	0.1645	0.1665	0.1655
FC	5.1050	−0.0355	0.0355	0.0304

mum absolute error is in the range of (0.0005, 0.0177), and the mean square error is in the range of (0.0003, 0.0112). The maximum absolute and the mean square errors of Gaussian noise are the smallest and those of the Lévy distribution noise are the largest, with the relationships expressed as

$$\text{Rerr}\,1(\text{FA}) < \text{Rerr}\,1(\text{FC}) < \text{Rerr}\,1(\text{FB}), \tag{3.15}$$

$$\text{Rerr}\,2(\text{FA}) < \text{Rerr}\,2(\text{FC}) < \text{Rerr}\,2(\text{FB}). \tag{3.16}$$

The corresponding fitting curves are depicted in Figures 3.5–3.9. We see that the results of Gaussian noise fitting have the highest accuracy, and the stretched Gaussian

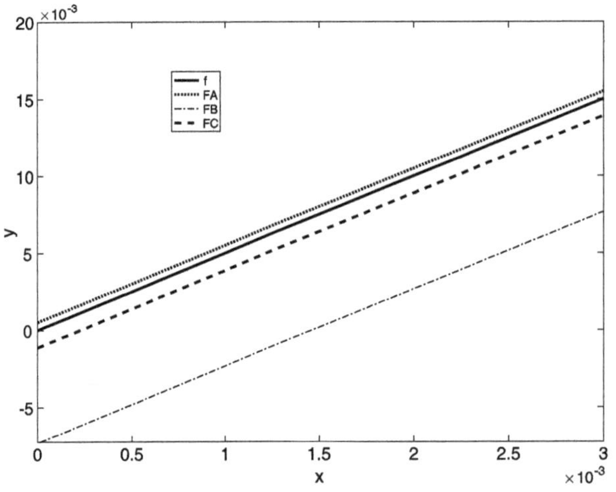

Figure 3.5: Linear model fitting to 1% noise.

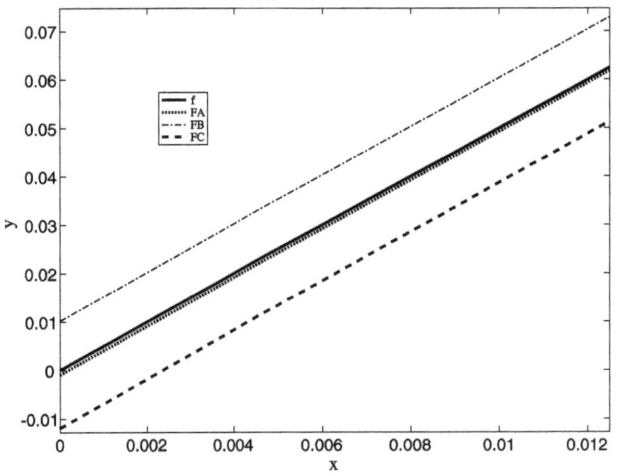

Figure 3.6: Linear model fitting to 5% noise.

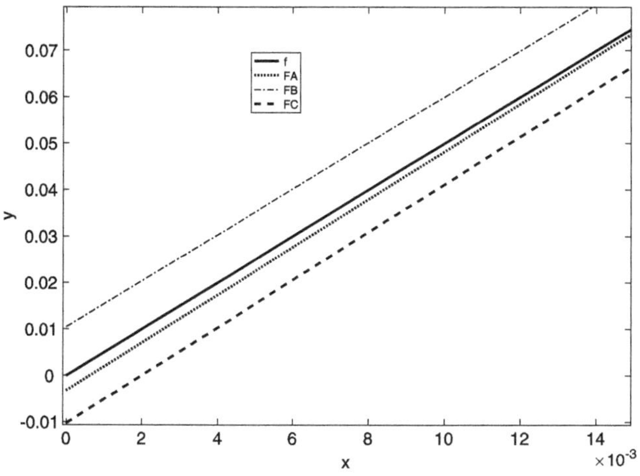

Figure 3.7: Linear model fitting to 10 % noise.

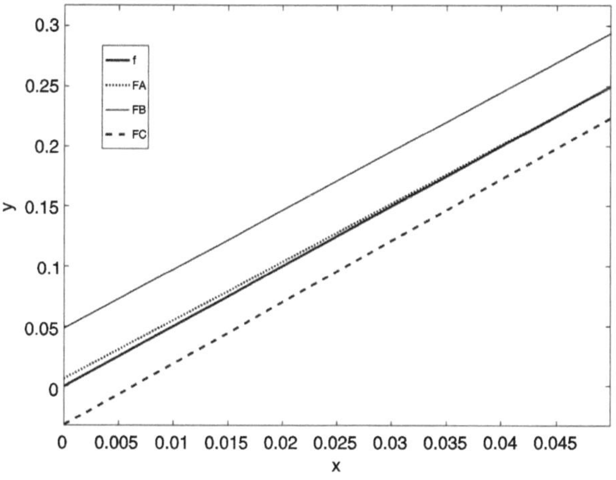

Figure 3.8: Linear model fitting to 15 % noise.

noise fitting curves are closer to those of the Gaussian noise compared with the results of Lévy noise data fitting.

b) A polynomial can be constructed by means of addition, multiplication, and exponentiation to a non-negative power, which is usually written as the following form with a single variable x:

$$f(x) = a_n x^n + a_{n-1} x^{n-1} + \cdots + a_2 x^2 + a_1 x + a_0, \tag{3.17}$$

where $a_0, a_1, \ldots, a_{n-1}, a_n$ are constants. We select three parameters of the following polynomial function:

$$F(x) = ax^2 + bx + c \quad (a \neq 0). \tag{3.18}$$

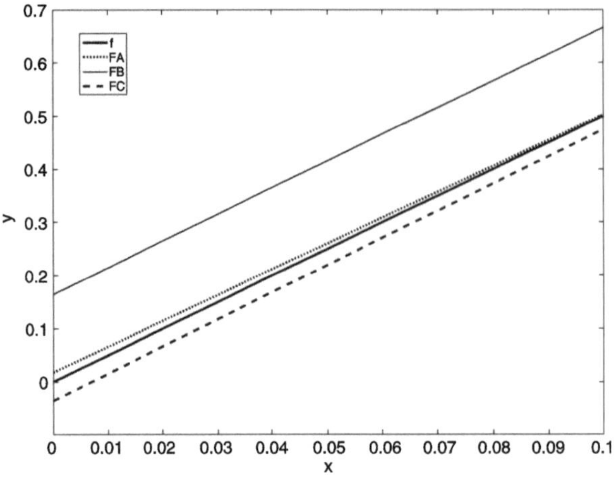

Figure 3.9: Linear model fitting to 20 % noise.

Here the following case is used as example:

$$y = 4x^2 + 2x + 3 \quad (a \neq 0). \tag{3.19}$$

Tables 3.6 to 3.10 give the estimated parameters and the errors for five different levels of noise in the polynomial case. The corresponding fitting curves are depicted in Figures 3.10–3.14. The Gaussian noise fitting maximum absolute error is in the range of (0.0019, 0.0170) and the mean square error is in the range of (0.0009, 0.0069).

Table 3.6: The estimated results for 1 % noise in the polynomial case.

function	*a*	*b*	*c*	Rerr1	Rerr2
f	4	2	3	0	0
FA	2.4931	2.0588	3.0013	0.0019	0.0009
FB	4.6313	1.9839	2.9963	0.0038	0.0036
FC	6.0044	1.8718	3.0035	0.0035	0.0020

Table 3.7: The estimated results for 5 % noise in the polynomial case.

function	*a*	*b*	*c*	Rerr1	Rerr2
f	4	2	3	0	0
FA	3.7673	2.0500	2.9974	0.0026	0.0010
FB	4.6455	1.8825	3.0166	0.0190	0.0137
FC	4.7226	1.8644	2.9985	0.0079	0.0058

Table 3.8: The estimated results for 10 % noise in the polynomial case.

function	*a*	*b*	*c*	Rerr1	Rerr2
f	4	2	3	0	0
FA	4.4244	1.8693	3.0071	0.0071	0.0030
FB	4.6928	1.8345	3.0367	0.0367	0.0295
FC	3.5281	2.1336	2.9754	0.0246	0.0178

Table 3.9: The estimated results for 15 % noise in the polynomial case.

function	*a*	*b*	*c*	Rerr1	Rerr2
f	4	2	3	0	0
FA	4.9474	1.7866	3.0070	0.0128	0.0048
FB	4.3098	1.9391	3.0338	0.0379	0.0327
FC	4.4846	1.9551	2.9692	0.0318	0.0270

Table 3.10: The estimated results for 20 % noise in the polynomial case.

function	*a*	*b*	*c*	Rerr1	Rerr2
f	4	2	3	0	0
FA	6.1102	1.6197	3.0086	0.0170	0.0069
FB	3.0272	2.2146	3.0432	0.0550	0.0517
FC	4.0612	2.0446	2.9506	0.0494	0.0443

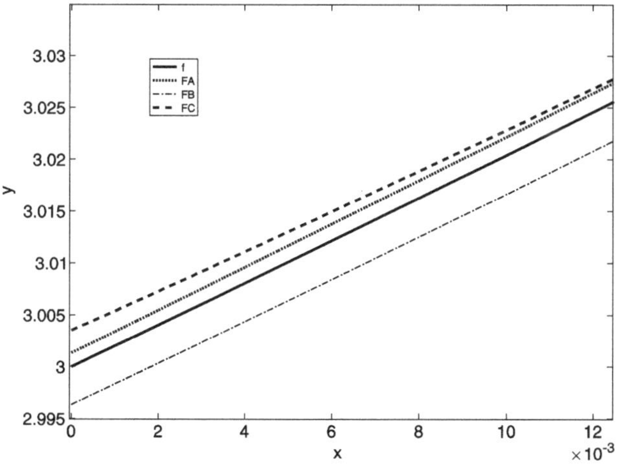

Figure 3.10: Polynomial model fitting to 1% noise.

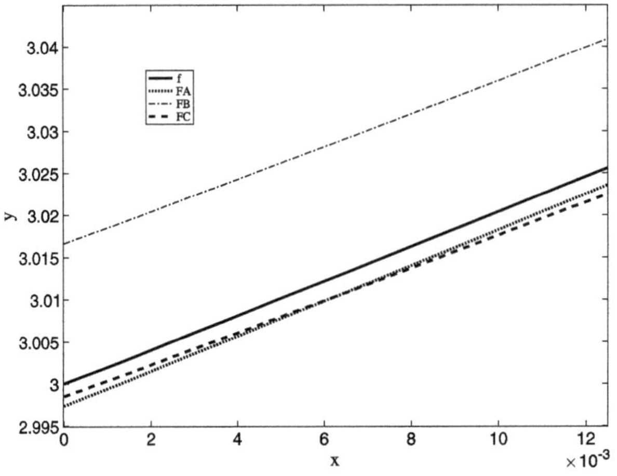

Figure 3.11: Polynomial model fitting to 5% noise.

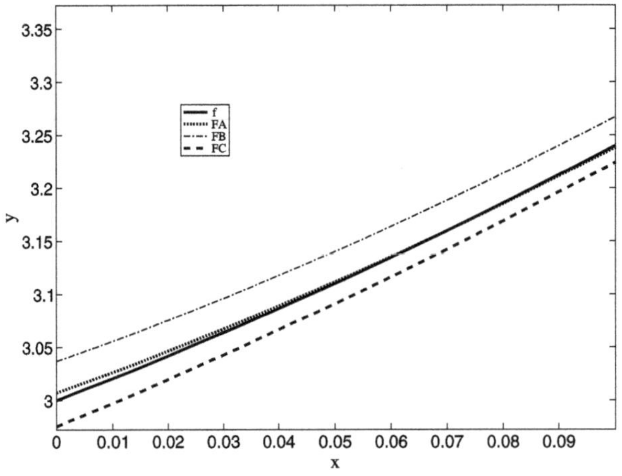

Figure 3.12: Polynomial model fitting to 10 % noise.

In Figures 3.10–3.13, we see the results of Gaussian noise fitting are the best, and the stretched Gaussian noise fitting curves are better than the results of Lévy noise data fitting.

c) Non-linear equations can be divided into two categories, i. e., polynomial equations and non-polynomial equations. In this part, we select the four parameters of the following exponential function:

$$F(x) = ae^{3x} + be^{-x} - cx + d \quad (a \neq 0). \tag{3.20}$$

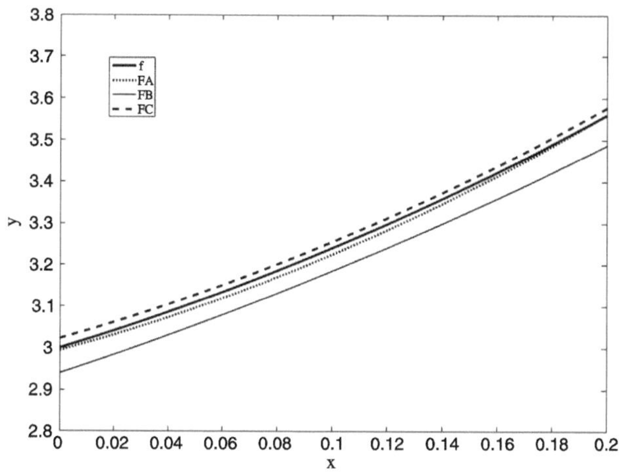

Figure 3.13: Polynomial model fitting to 15 % noise.

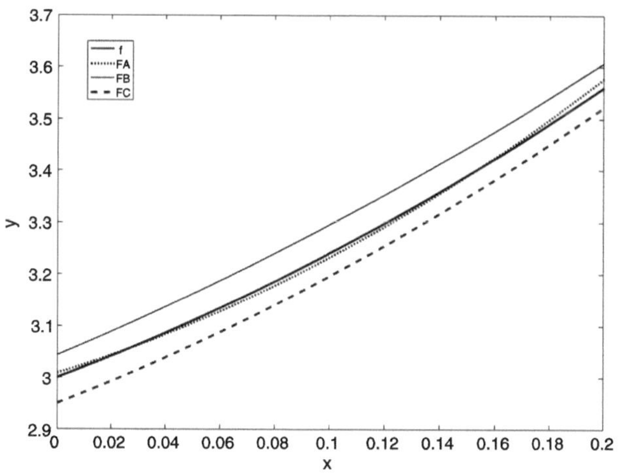

Figure 3.14: Polynomial model fitting to 20 % noise.

Here the following case is used as an example:

$$f(x) = 0.5e^{3x} + 0.2e^{-x} - x + \frac{1}{3}. \tag{3.21}$$

Tables 3.11 to 3.15 give the estimated parameters and the errors for five different levels of noise in the exponential function case. The corresponding fitting curves are shown in Figures 3.15–3.19. The Gaussian noise fitting data maximum absolute error is in the range of (0.0013, 0.0.137), the mean square error is in the range of (0.0004, 0.0075). The results of exponential cases have patterns similar to those shown in the linear and polynomial cases.

Table 3.11: The estimated results for 1 % noise in the exponential case.

function	*a*	*b*	*c*	*d*	Rerr1	Rerr2
f	0.5	0.2	1	1/3	0	0
FA	0.4651	0.8096	0.3043	−0.2412	0.0013	0.0004
FB	0.4887	0.4365	0.7425	0.1168	0.0087	0.0081
FC	0.4889	0.3664	0.8068	0.1795	0.0015	0.0009

Table 3.12: The estimated results for 5 % noise in the exponential case.

function	*a*	*b*	*c*	*d*	Rerr1	Rerr2
f	0.5	0.2	1	1/3	0	0
FA	0.4282	1.5436	−0.5101	−0.9386	0.002	0.0012
FB	0.5493	−0.7494	2.0662	1.2535	0.0211	0.0198
FC	0.6557	−3.1130	4.5957	3.4751	0.0156	0.0078

Table 3.13: The estimated results for 10 % noise in the exponential case.

function	*a*	*b*	*c*	*d*	Rerr1	Rerr2
f	0.5	0.2	1	1/3	0	0
FA	0.7448	−4.4825	6.2056	4.7692	0.0137	0.0075
FB	0.3697	2.9448	−1.9627	−2.3200	0.0545	0.0496
FC	0.8745	−6.9997	9.0002	7.1302	0.0284	0.0158

Table 3.14: The estimated results for 15 % noise in the exponential case.

function	*a*	*b*	*c*	*d*	Rerr1	Rerr2
f	0.5	0.2	1	1/3	0	0
FA	0.1986	5.1211	−4.7226	−4.2859	0.0097	0.0038
FB	0.5151	−0.6695	1.8449	1.1120	0.0867	0.0763
FC	0.4250	1.1082	−0.1989	−0.5301	0.0303	0.0206

Table 3.15: The estimated results for 20 % noise in the exponential case.

function	*a*	*b*	*c*	*d*	Rerr1	Rerr2
f	0.5	0.2	1	1/3	0	0
FA	0.3164	3.9831	−3.1994	−3.2706	0.0091	0.0063
FB	0.3339	3.7390	−2.8727	−2.9547	0.0873	0.0823
FC	0.6920	−3.3436	4.9695	3.6399	0.0451	0.0371

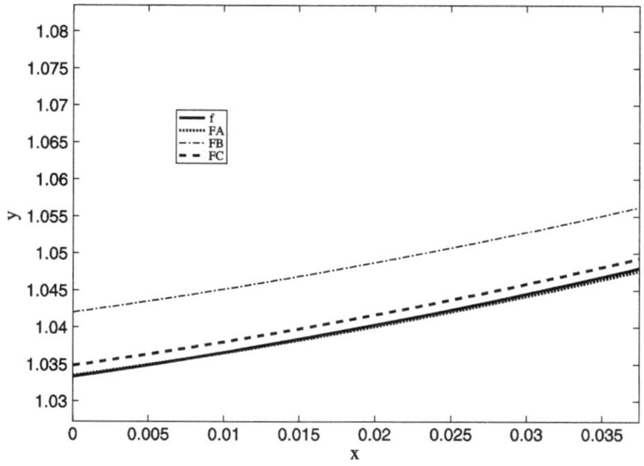

Figure 3.15: Exponential model fitting to 1% noise data.

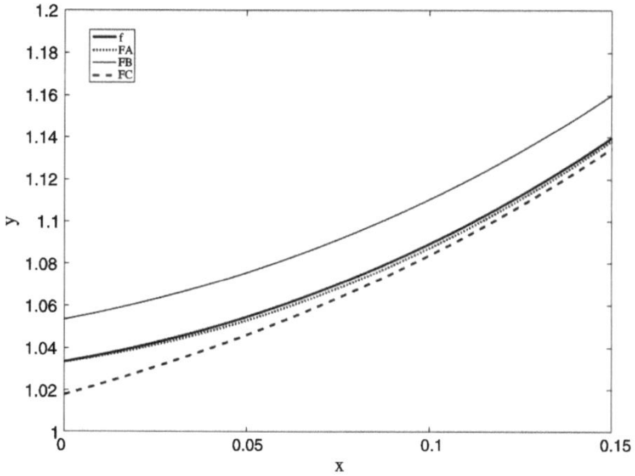

Figure 3.16: Exponential model fitting to 5% noise data.

To summarize all the above results, we found that the maximum absolute and the mean square errors for the Gaussian noise cases are the smallest, but the values for the Lévy noise cases are the largest, i. e.,

$$\text{Rerr}\,1(FA) < \text{Rerr}\,1(FC) < \text{Rerr}\,1(FB), \tag{3.22}$$

$$\text{Rerr}\,2(FA) < \text{Rerr}\,2(FC) < \text{Rerr}\,2(FB). \tag{3.23}$$

It can be observed from Figures 3.5 to 3.19 that the results of Gaussian noise fitting have the best accuracy, and the stretched Gaussian noise fitting curves are closer to those of the Gaussian noise compared with the results of Lévy noise data fitting. Thus, the least square method is less accurate when it is applied to the non-Gaussian noise data fitting compared with the cases of Gaussian noise, especially when the noise level is larger than 5 %.

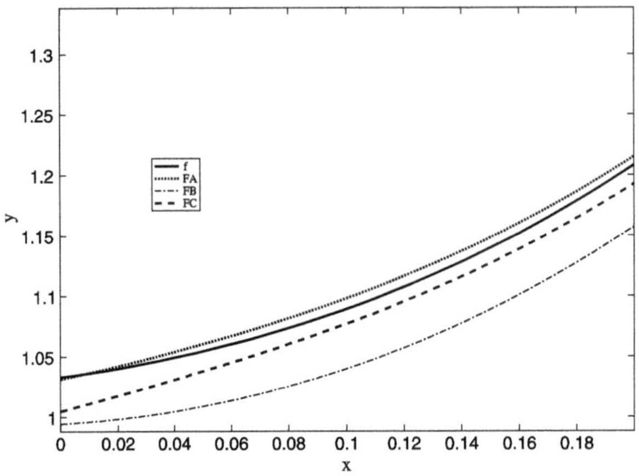

Figure 3.17: Exponential model fitting to 10 % noise.

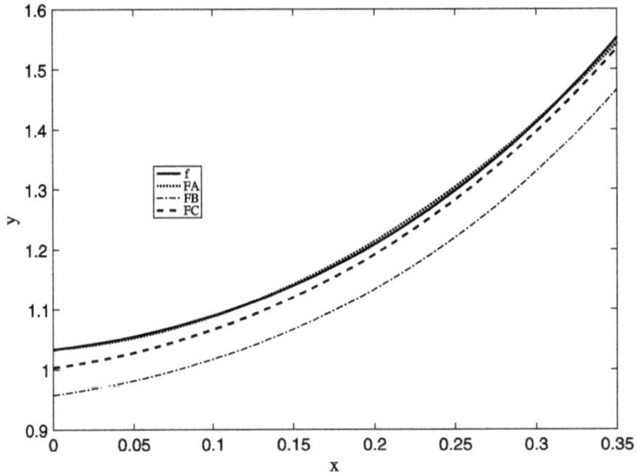

Figure 3.18: Exponential model fitting to 15 % noise.

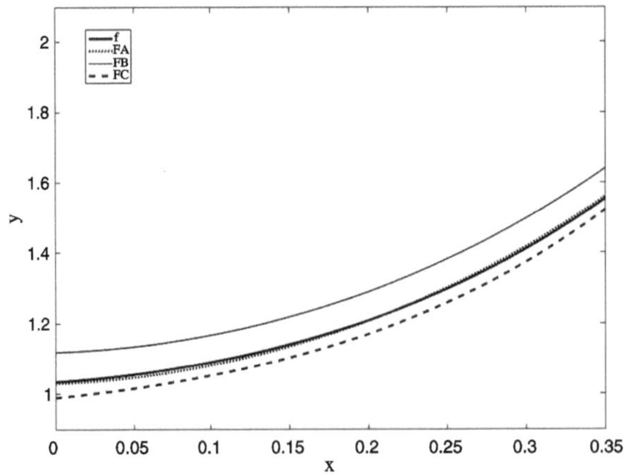

Figure 3.19: Exponential model fitting to 20 % noise.

4 Lyapunov stability of Hausdorff derivative non-linear systems

4.1 Fractal time

First, we introduce the concept of fractal time, which is essential in the formulation of Hausdorff derivatives. The fractal time τ_α is a measurement of time which has fractal dimensionality $0 < \alpha < 1$. Mathematically, the relation of the fractal time τ_α and the absolute time t can be expressed as

$$\Delta\tau_\alpha = \Delta(t^\alpha). \tag{4.1}$$

The non-linearity of the fractal time is illustrated in Figure 4.1.

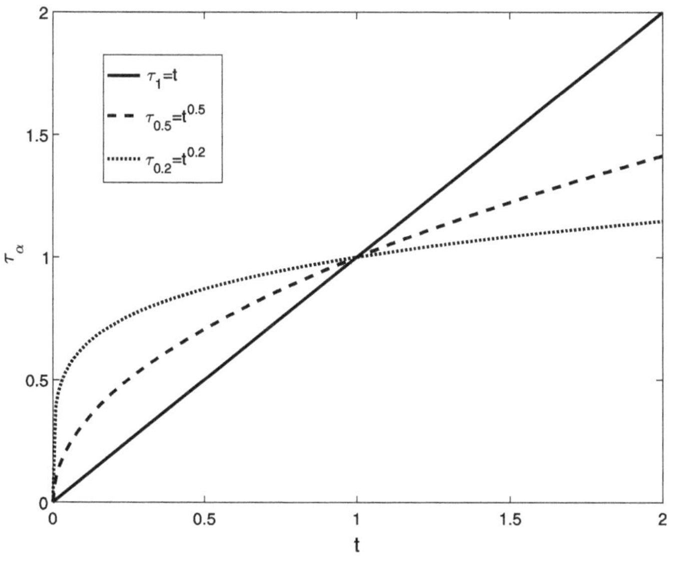

Figure 4.1: Fractal time τ_α ($\alpha = 0.2, 0.5, 1$) as a function of the absolute time t.

Figure 4.1 shows that the "fractal time clock" $\tau_{0.2}$, $\tau_{0.5}$ goes much faster than the "absolute time clock" τ_1 when the absolute time is not far from zero. More precisely, if we set $t_0 = (\frac{1}{\alpha})^{\frac{1}{\alpha-1}}$, it can be calculated that when $0 < t < t_0$, the "fractal clock" τ_α always goes faster than the "absolute time clock". But after t_0, the "fractal time clock" will go slower. This conclusion is also illustrated by Figure 4.2, where the time scales of $\tau_\alpha(\alpha = 0.2, 0.5, 1)$ are plotted. From Figure 4.2, it is easy to conclude that the time scale of fractal time τ_α should be a fractal with dimensionality α when $\Delta t \to 0$.

https://doi.org/10.1515/9783110608526-004

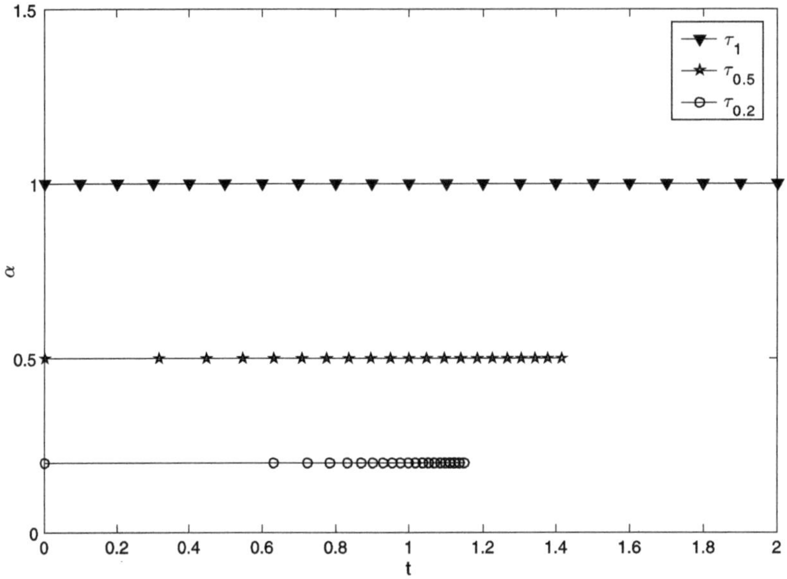

Figure 4.2: Comparison of the time scales of fractal time τ_α ($\alpha = 0.2, 0.5, 1$).

Considering a particle traveling in fractal time with a velocity $v(\tau_\alpha)$, $\tau_\alpha \in [0, T^\alpha]$, $T > 0$ is a constant time. When the velocity is constant, the travel distance l of this particle as a function of absolute time $\tau_1 \in [0, T]$ is expressed as

$$l(\tau_\alpha) = v\tau_\alpha = v\tau_1^\alpha. \tag{4.2}$$

When the velocity v varies in time, the travel distance as a function of absolute time t can be expressed by the Hausdorff integral as follows:

$$l(t) = \int_0^t v(\tau_1^\alpha)d\tau_1^\alpha = \int_0^t \alpha v(\tau_1^\alpha)\tau_1^{\alpha-1}d\tau_1. \tag{4.3}$$

Applying the first-order derivative with respect to t on both sides of equation (4.3), we have

$$\frac{dl(t)}{dt} = \alpha v(t^\alpha)t^{\alpha-1}. \tag{4.4}$$

Hence the velocity of the particle can be expressed as

$$v(t^\alpha) = \frac{dl(t)}{\alpha t^{\alpha-1}dt} = \frac{dl(t)}{d(t^\alpha)} = \lim_{t'\to t}\frac{l(t') - l(t)}{(t')^\alpha - t^\alpha}, \tag{4.5}$$

which can be used to measure the velocity of physical processes in term of fractal time. The above expression motivates us to define the Hausdorff derivative [31] of a function

$u(t)$ as follows:

$$D^{\alpha}u(t) = \frac{du(t)}{dt^{\alpha}} = \lim_{t' \to t} \frac{u(t') - u(t)}{t'^{\alpha} - t^{\alpha}}, \tag{4.6}$$

which can be used to measure the velocity of physical processes in term of fractal time. From equation (4.6), we have $D^{\alpha}u(t) = \frac{du(t)}{dt^{\alpha}} = \frac{du(t)}{\alpha t^{\alpha-1}dt}$. This relation shows that we can use the standard numerical techniques for the integer-order derivative equations for obtaining the approximate solutions of Hausdorff dynamical systems.

4.2 Hausdorff dynamical systems and the concept of stretched exponential stability

In this section, we will apply the direct Lyapunov method to investigate the non-linear Hausdorff dynamical systems with time-varying delay as follows:

$$\begin{cases} D^{\alpha}x(t) = f(x(t), t), & t > 0, \\ x(0) = x_0, & x_0 \in \aleph, \end{cases} \tag{4.7}$$

where $x(t) \in \aleph$ is the state vector, $\aleph \subseteq R^n$ is a domain containing the origin $x = 0$, and $f : \aleph \times [0, \infty) \to R^n$ is a non-linear piecewise continuous function in t.

Without loss of generality, we assume that the zero solution $x = 0$ is an equilibrium solution of equation system (4.7), i. e., $D^{\alpha}(0) = f(0, t) \equiv 0$. It is well known that the fractional derivative diffusion equation underlies the Mittag-Leffler function decaying pattern, which leads to the Mittag-Leffler stability [112, 113].

Definition 4.1 (Mittag-Leffler stability). The zero solution of a dynamical system is said to be Mittag-Leffler stable if the non-zero solution $x(t)$ satisfies

$$\|x(t)\| \le \{m[x(0)]E_{\alpha}(-\lambda t^{\alpha})\}^{\beta}, \tag{4.8}$$

where $E_{\alpha}(\cdot)$ is the Mittag-Leffler function, $0 < \alpha < 1$, $\beta > 0$, $\lambda > 0$, function $m(x)$ is locally Lipschitz on $x(0) \in R^n$ with Lipschitz constant m_0, and $m(0) = 0$, $m(x) \ge 0$.

In contrast, the Hausdorff diffusion equation characterizes the fractal exponential decay, also known as the stretched relaxation and the Kohlrausch–Williams–Watts (KWW) stretched Gaussian [98]. Hence, by analogy, we propose the definition of stretched exponential stability for the Hausdorff differential systems.

Definition 4.2 (Stretched exponential stability). The zero solution of equation system (4.7) is stretched exponentially stable if

$$\|x(t)\| \le \{m[x(0)] \exp(-\gamma t^{\alpha})\}^{\beta}, \tag{4.9}$$

where $0 < \alpha \le 1$, $\gamma, \beta > 0$, function $m(x)$ is locally Lipschitz on $x \in \aleph$ with Lipschitz constant m_0, and $m(0) = 0$, $m(x) \ge 0$.

From Definition 4.2, it is clear that the Mittag-Leffler stability and the stretched exponential stability both include the exponential stability as a special case when $\alpha = 1$. However, they are controlled by different decaying patterns when $\alpha \neq 1$, as shown in Figure 4.3.

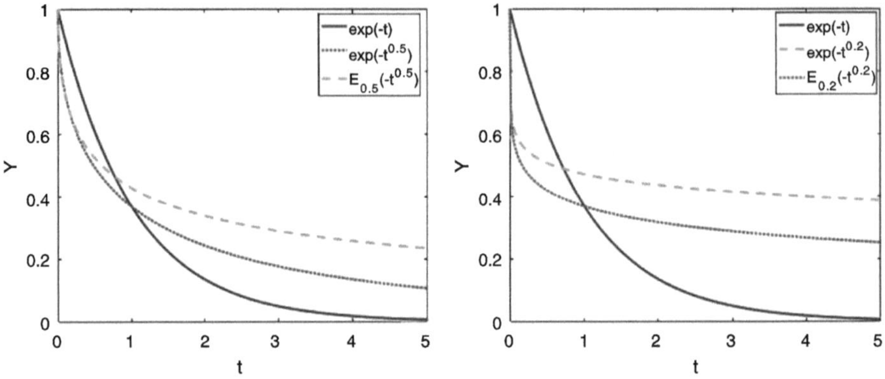

Figure 4.3: Comparison of decaying patterns of the stretched exponential functions and the Mittag-Leffler functions.

Next we prove that the stretched exponential stability implies Lyapunov stability.

Lemma 4.1. *If the zero solution of equation system* (4.7) *is stretched exponential stable, then it is also Lyapunov asymptotically stable.*

Proof. (i) For every $\varepsilon > 0$, set $\delta = \frac{\varepsilon^{\frac{1}{\beta}}}{m_0}$. Then if $\|x(0)\| < \delta$, from Definition 4.2, $\|x(t)\| \leq \{m[x(0)] \exp(-\gamma t^{\alpha})\}^{\beta} < (m_0\delta)^{\beta} = \varepsilon$, for every $t > 0$.

(ii) From Definition 4.2, we have $\lim_{t\to+\infty} \|x(t)\| \leq \lim_{t\to+\infty}\{m[x(0)] \exp(-\gamma t^{\alpha})\}^{\beta}=0$.

From (i) and (ii), we conclude that the zero solution of equation system (4.7) is Lyapunov asymptotically stable. □

4.3 Application of the Lyapunov direct method

For non-linear dynamical systems, the Lyapunov direct method provides an efficient way to establish the stability criteria without explicitly solving the equations. The main idea of the Lyapunov direct method is to find a Lyapunov function satisfying certain conditions for the system. If such function exists, then we can conclude that the system is stable. It should be pointed out that the criteria obtained by the Lyapunov direct method are sufficient, which means that if one cannot find the right Lyapunov function, the system may still be stable.

Now we apply the Lyapunov direct method to the case of dynamical systems based on the Hausdorff derivative, which leads to the stretched exponential stability.

Theorem 4.1. *Let $V(x,t) : \aleph \times [0,\infty] \to R$ be a continuously differentiable function and*
(i) *let it be locally Lipschitz with respect to x such that*

$$c_1\|x\|^a \le V(x,t) \le c_2\|x\|^b, \tag{4.10}$$

(ii) *we have*

$$D^\beta V(x,t) \le -c_3\|x\|^b, \tag{4.11}$$

where $0 < \beta \le 1$, c_i $(i = 1,2,3)$, a, b are positive constants. Then the zero solution of equation system (4.7) is stretched exponentially stable.

Proof. From conditions (i) and (ii), we have

$$D^\beta V(x,t) \le -\frac{c_3}{c_2}V(x,t). \tag{4.12}$$

Hence there exists a non-negative function $m(t)$ satisfying

$$D^\beta V(x,t) + m(t) = -\frac{c_3}{c_2}V(x,t). \tag{4.13}$$

Applying the variable transform

$$t = \hat{t}^{\frac{1}{\beta}}, \tag{4.14}$$

we have $D^\beta V(x,t) = \frac{d}{d\hat{t}}V(x(\hat{t}^{\frac{1}{\beta}}), \hat{t}^{\frac{1}{\beta}})$, and equation (4.13) becomes

$$\frac{d}{d\hat{t}}V(x(\hat{t}^{\frac{1}{\beta}}), \hat{t}^{\frac{1}{\beta}}) + m(\hat{t}^{\frac{1}{\beta}}) = -\frac{c_3}{c_2}V(x(\hat{t}^{\frac{1}{\beta}}), \hat{t}^{\frac{1}{\beta}}). \tag{4.15}$$

Let $\underline{V}(\hat{t}) = V(x(\hat{t}^{\frac{1}{\beta}}), \hat{t}^{\frac{1}{\beta}})$, $\underline{m}(\hat{t}) = m(\hat{t}^{\frac{1}{\beta}})$. Then equation (4.15) becomes

$$\frac{d}{d\hat{t}}\underline{V}(\hat{t}) + \underline{m}(\hat{t}) = -\frac{c_3}{c_2}\underline{V}(\hat{t}). \tag{4.16}$$

Taking the Laplace transform $\mathcal{L}$ on both sides of equation (4.16) gives

$$s\bar{V}(s) - V(x(0),0) + \bar{m}(s) = -\frac{c_3}{c_2}\bar{V}(s), \tag{4.17}$$

where $\bar{V}(s) = \mathcal{L}\underline{V}(\widehat{t})$, $\bar{m}(s) = \mathcal{L}\underline{m}(\widehat{t})$.

It follows that

$$\bar{V}(s) = \frac{V(x(0),0) - \bar{m}(s)}{s + \frac{c_3}{c_2}}. \tag{4.18}$$

Taking the inverse Laplace transform of equation (4.18) yields

$$\underline{V}(\hat{t}) = V(x(0), 0) \exp\left(-\frac{c_3}{c_2}\hat{t}\right) - \underline{m}(\hat{t}) * \exp\left(-\frac{c_3}{c_2}\hat{t}\right), \tag{4.19}$$

where $*$ denotes the convolution operator.

Applying the following variable transform on equation (4.19),

$$\hat{t} = t^{\beta}, \tag{4.20}$$

it follows that

$$V(x(t), t) = V(x(0), 0) \exp\left(-\frac{c_3}{c_2}t^{\beta}\right) - m(t) * \exp\left(-\frac{c_3}{c_2}t^{\beta}\right). \tag{4.21}$$

From equation (4.15), we get

$$V(x(t), t) \le V(x(0), 0) \exp\left(-\frac{c_3}{c_2}t^{\beta}\right), \tag{4.22}$$

since $m(t) * \exp(-\frac{c_3}{c_2}t^{\beta}) \ge 0$.

Based on equation (4.22) together with equation (4.10), we have

$$c_1 \|x(t)\|^{a} \le V(x(0), 0) \exp\left(-\frac{c_3}{c_2}t^{\beta}\right), \tag{4.23}$$

which means that

$$\|x(t)\| \le \left[\frac{V(x(0), 0)}{c_1} \exp\left(-\frac{c_3}{c_2}t^{\beta}\right)\right]^{\frac{1}{a}}. \tag{4.24}$$

□

Next, the fractal comparison principle is applied to investigate stability conditions with class-κ functions.

Lemma 4.2 (Fractal comparison principle). *Let $x(t) : [0, \infty) \to [0, \infty)$, $y(t) : [0, \infty) \to [0, \infty)$ be differentiable continuous functions of t satisfying $x(0) = y(0)$, $D^{\alpha}x(t) \ge D^{\alpha}y(t)$, $\alpha \in (0, 1]$. Then we get $x(t) \ge y(t)$ for $t \ge 0$.*

Proof. From $D^{\alpha}x(t) \ge D^{\alpha}y(t)$, $\alpha \in (0, 1]$, there exists a non-negative function $m(t)$ such that

$$D^{\alpha}x(t) = m(t) + D^{\alpha}y(t). \tag{4.25}$$

Applying the variable transform $t = \hat{t}^{\frac{1}{\alpha}}$ to equation (4.25), it becomes

$$\frac{d}{d\hat{t}}x(\hat{t}^{\frac{1}{\alpha}}) = m(\hat{t}^{\frac{1}{\alpha}}) + \frac{d}{d\hat{t}}y(\hat{t}^{\frac{1}{\alpha}}). \tag{4.26}$$

Let $\underline{x}(\hat{t}) = x(\hat{t}^{\frac{1}{\alpha}})$, $\underline{y}(\hat{t}) = x(\hat{t}^{\frac{1}{\alpha}})$, $\underline{m}(\hat{t}) = m(\hat{t}^{\frac{1}{\alpha}})$. Then the above equation becomes

$$\frac{d}{d\hat{t}}\underline{x}(\hat{t}) = \underline{m}(\hat{t}) + \frac{d}{d\hat{t}}\underline{y}(\hat{t}). \tag{4.27}$$

Taking the Laplace transform on both sides of equation (4.27) gives

$$[s\bar{x}(s) - x(0)] = \bar{m}(s) + [s\bar{y}(s) - y(0)], \tag{4.28}$$

where $\bar{x}(s) = \mathcal{L}\underline{x}(\hat{t})$, $\bar{y}(s) = \mathcal{L}\underline{y}(\hat{t})$.

From $x(0) = y(0)$, equation (4.28) reduces to

$$\bar{x}(s) = \frac{\bar{m}(s)}{s} + \bar{y}(s). \tag{4.29}$$

Applying the inverse Laplace transform on both sides of equation (4.29), it gives

$$\underline{x}(\hat{t}) = \underline{m}(\hat{t}) * 1 + \underline{y}(\hat{t}). \tag{4.30}$$

Then we apply the variable transform $\hat{t} = t^{\alpha}$ to both sides of equation (4.30); we have

$$x(t) = \int_0^{t^{\alpha}} m(\tau)d\tau + y(t). \tag{4.31}$$

It follows from $m(t) \geq 0$ that

$$x(t) \geq y(t), \quad t \geq 0. \tag{4.32}$$

□

Definition 4.3. A continuous function $k(t) : [0,\infty) \to [0,\infty)$ is said to be a class-κ function if it is strictly increasing and $k(0) = 0$ [102].

Theorem 4.2. *Assuming that there exists a Lyapunov function $V(x,t)$ and a class-κ function k_i $(i = 1,2,3)$ such that*

$$k_1(\|x\|) \leq V(x,t) \leq k_2(\|x\|), \tag{4.33}$$

$$D^{\beta}V(x,t) \leq -k_3(\|x\|), \tag{4.34}$$

where $\beta \in (0,1]$, the zero solution of equation system (4.7) is stretched exponentially stable.

Proof. From equation (4.33) and equation (4.34), we have

$$D^{\beta}V(x,t) \leq -k_3(k_2^{-1}(V(x,t))) \leq 0 = D^{\beta}(V(x(0),0)). \tag{4.35}$$

By using Lemma 4.2 and $V(x,t) \geq 0$, we have $V(x,t) \leq V(x(0),0)$.

Case 4.1. If there exists $t_1 \geq 0$, such that $V(x(t_1), t_1) = 0$, it follows from equation (4.33) that $x(t_1) = 0$. Then we have $x = 0$, for $x \geq t_1$, since $x = 0$ is an equilibrium solution of equation system (4.7).

Case 4.2. Suppose there exists a positive constant $\varepsilon > 0$ such that $V(x,t) \geq \varepsilon$ for $t \in [0, +\infty)$. Then

$$0 < \varepsilon \leq V(x,t) \leq V(x(0),0), \quad t \in [0,+\infty). \tag{4.36}$$

From (4.35) together with (4.36), we have

$$\begin{aligned} D^{\beta}V(x,t) &\leq -k_3(k_2^{-1}(V(x,t))) \leq -k_3(k_2^{-1}(\varepsilon)) = -\frac{k_3(k_2^{-1}(\varepsilon))}{V(x(0),0)} V(x(0),0) \\ &\leq -\frac{k_3(k_2^{-1}(\varepsilon))}{V(x(0),0)} V(x,t). \end{aligned} \tag{4.37}$$

By employing the same proof method used in Theorem 4.1, we have $\lim_{t\to\infty} V(x,t) = 0$, which contradicts with $V(x,t) \geq \varepsilon$ for $t \in [0,+\infty)$.

Based on the results in Case 4.1 and Case 4.2, we can conclude that $\lim_{t\to\infty} V(x,t) = 0$ and $V(x(t),t) \leq V(x(0),0) \exp(-\frac{k_3(k_2^{-1}(\varepsilon))}{V(x(0),0)} t^{\beta})$.

Then from equation (4.33) we conclude the zero solution of equation system (4.7) is stretched exponentially stable. □

The comparison between the Hausdorff derivative and the classical fractional derivative for modeling anomalous diffusion has already been reported in [45], where the simplicity and efficiency of the Hausdorff derivative is illustrated. In this section, two examples are presented to illustrate the stability. Moreover, necessary numerical comparisons of the stability behavior of the Hausdorff derivative system and its corresponding classical fractional-order (in Caputo sense) system will be presented in this section.

Example 4.1. We have the one-dimensional dynamical system

$$D^{\alpha}x(t) = -x(t)\cos^2[x(t)], \tag{4.38}$$

where $\alpha \in (0,1)$, $x(t) : [0,\infty) \to R$. We claim that the zero solution of equation (4.38) is stretched exponentially stable.

Proof. Let $V(x(t),t) = x^2(t)$. Then

$$D^{\alpha}[V(x(t),t)] = D^{\alpha}[x^2(t)] = -2x^2(t)\cos^2[x(t)] \leq -2x^2(t). \tag{4.39}$$

Then by Theorem 4.1, the zero solution of equation (4.38) is stretched exponentially stable. For $\alpha = 0.5, 0.7, 0.9$ and $x(0) = 1$, the solutions of equation system (4.38) are plotted in Figure 4.4.

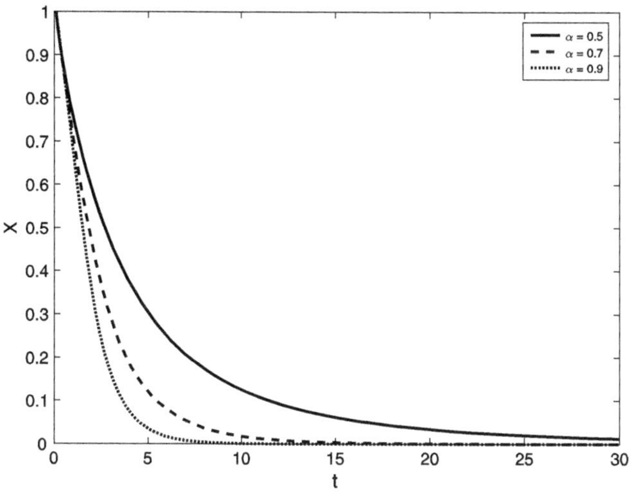

Figure 4.4: The solution of system (4.38) with $x(0) = 1$ for $\alpha = 0.5, 0.7, 0.9$.

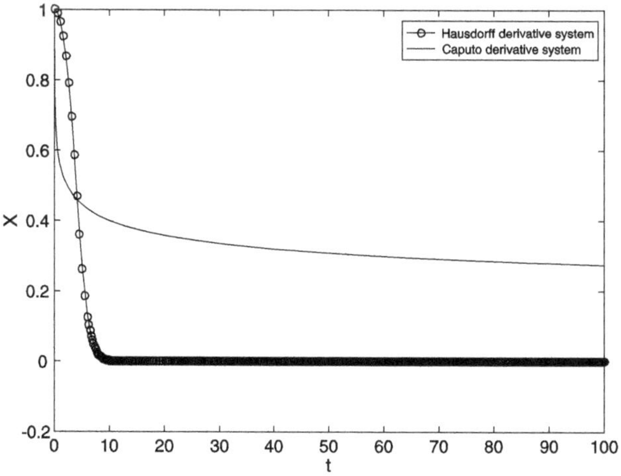

Figure 4.5: Comparisons of numerical solutions of Hausdorff dynamical systems and fractional-order systems when $x(0) = 1$, $\alpha = 0.2$.

A comparison of the numerical solutions of equation system (4.38) and its corresponding systems when replacing the Hausdorff derivative D^{α} by the Caputo derivative ${}_0^C D_t^{\alpha}$ is presented in Figures 4.5 and 4.6. The numerical solutions of Hausdorff dynamical systems can be obtained by the function ODE45 in *MATLAB*®; the numerical solutions of fractional-order (Caputo sense) systems can be obtained by the function “fde12” created by Roberto Garrappa [72].

It is indicated in Figures 4.5 and 4.6 that the Hausdorff derivative systems converge to zero much faster than its corresponding fractional-order system. □

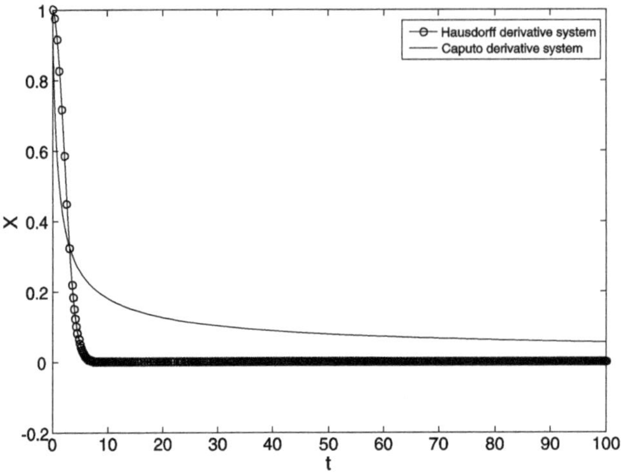

Figure 4.6: Comparison of numerical solutions of Hausdorff dynamical systems and fractional-order systems when $x(0) = 1$, $\alpha = 0.5$.

Example 4.2. Consider the following three dimensional dynamic system:

$$\begin{cases} D^\alpha x_1(t) = -x_1(t), \\ D^\alpha x_2(t) = -x_2(t), \\ D^\alpha x_3(t) = -x_3(t) - x_3(t)^3, \end{cases} \tag{4.40}$$

where $\alpha = 0.5$.

Proof. Choose the Lyapunov function as $V(x(t), t) = x_1^2(t) + x_2^2(t) + x_3^2(t)$. Then

$$\begin{aligned} D^\alpha[V(x(t), t)] &= 2x_1(t)D^\alpha x_1(t) + 2x_2(t)D^\alpha x_2(t) + 2x_3(t)D^\alpha x_3(t) \\ &= -2x_1^2(t) - 2x_1^2(t) - 6x_1^4(t) \\ &\le -x_1^2(t) - x_2^2(t) - x_3^2(t). \end{aligned} \tag{4.41}$$

Then by Theorem 4.2, the zero solution of (4.40) is asymptotically stable. □

The convergence of the solution of equation system (4.40) is illustrated in Figure 4.7, where its numerical solution is obtained by the function ODE45 in MATLAB.

In this chapter, the stability of non-linear Hausdorff dynamical systems was investigated. By using the Lyapunov direct method, we introduced the concept of the stretched exponential stability, which is a Kullback–Leibler (KL)-type of bound on the state trajectories. The classical exponential stability is a special limiting case of the proposed stretched exponential stability when the stretch parameter equals 1. In addition, a fractal comparison principle is suggested to obtain stability conditions. Two examples are tested to examine the applicability of the proposed concept. Through

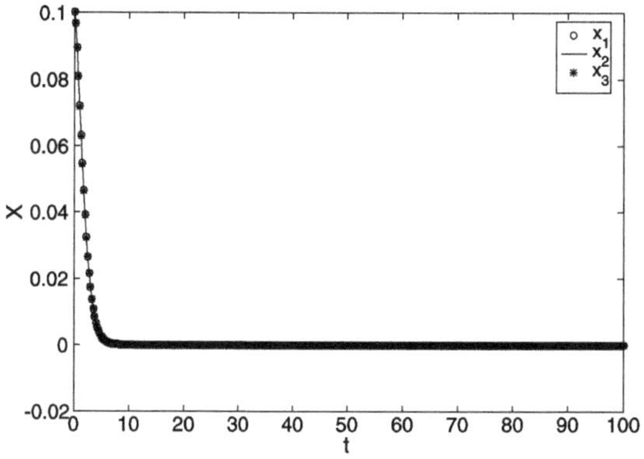

Figure 4.7: Numerical solution of equation system (4.40) with initial condition $x_1(0) = x_2(0) = x_3(0) = 0.1$.

comparison, it is indicated that the stretched exponential stability can serve as an alternative approach for characterizing the decaying patterns of the non-integer-order systems, and can find application fields different from the Mittag-Leffler stability. More work along this line is in progress. More specifically, our future works will consider applications of the proposed stability theory in synchronization of complex dynamical networks with time delay and impulsive effect [188, 189].

5 Hausdorff radial basis function

5.1 Singular boundary method using the fundamental solution

5.1.1 Singular boundary method for Hausdorff Laplacian

The singular boundary method (SBM) [33, 48] applies the fundamental solution of the Hausdorff Laplacian as the interpolation basis function to approximate the solution of the considered problems, in which the source and collocation points coincide on the physical boundary. The approximate representation of the SBM is expressed in terms of the linear combination of the fundamental solutions

$$u(\boldsymbol{x}^i) = \sum_{j=1,j\neq i}^{N} \alpha_j u^*(r^\beta(\boldsymbol{x}^i, \boldsymbol{y}^j)) + \alpha_i u_{ii}, \quad \boldsymbol{x}^i \in \Omega, \tag{5.1}$$

where $u^*(r^\beta(\boldsymbol{x}^i, \boldsymbol{y}^j))$ is the fundamental solution of the Hausdorff Laplacian, as shown in Table 2.1, $\boldsymbol{x}^i$ denotes the ith collocation point, $\boldsymbol{y}^j$ represents the jth source point, α_j is the jth unknown coefficient of the distributed sources, N is the number of boundary source points, and u_{ii} are called the origin intensity factors (OIFs), i. e., the diagonal elements of the SBM interpolation matrix.

The key issue of the SBM is to accurately evaluate the OIFs. There are four techniques available for determining the OIFs: the inverse interpolation technique; the subtracting and adding-back desingularization technique; the integral mean value approach; and the empirical formulas. Here we employ the integral average technique [44, 47] to determine the OIFs in the SBM, namely,

$$u_{ii} = \frac{1}{\ell_i} \int_{\Gamma_i} u^*(r^\beta(\boldsymbol{x}^i, \boldsymbol{y}))d\Gamma_{\boldsymbol{y}}, \tag{5.2}$$

where ℓ_i is the length or area of Γ_i and denotes the characteristic length of local influence domain Γ_i of the source point $\boldsymbol{x}^i$ as shown in Figure 5.1.

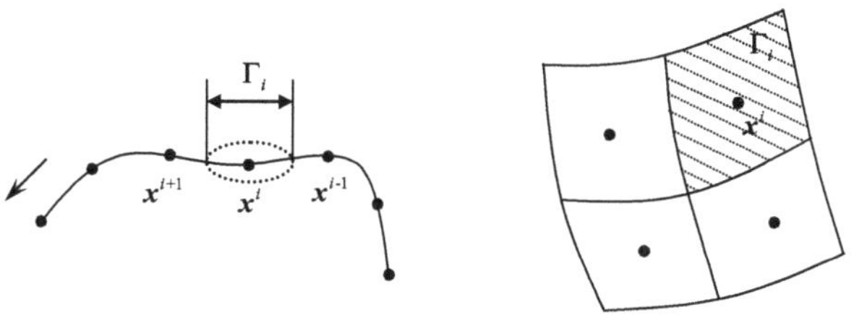

Figure 5.1: Nodal integration domain in (a) two-dimensional and (b) three-dimensional problems.

https://doi.org/10.1515/9783110608526-005

5.1.2 Numerical experiments

Example 5.1. The first case considers a two-dimensional heat transfer problem in a square medium $\Omega = \{(x, y) \mid 1 \le x, y \le 3\}$, as shown in Figure 5.2. The temperature on the boundary is assumed known and is given by

$$u(x, y) = e^x \sin(y) + e^y \sin(x). \tag{5.3}$$

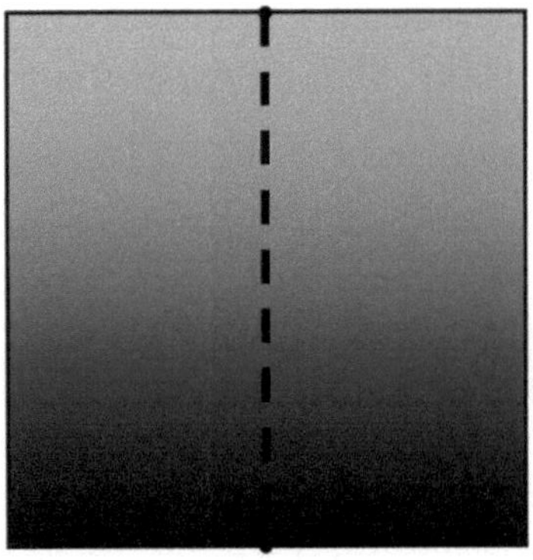

Figure 5.2: Steady heat transfer in a two-dimensional fractal medium.

The boundary element method (BEM) [51, 185, 186] is a powerful and versatile numerical technique for the numerical solution of certain boundary value problems whose fundamental solutions are available. The main advantage is its ability in reducing the dimensionality of a problem by one. However, the calculation of integration is troublesome and time consuming. To compare the SBM with the BEM in the integer-dimensional case, Table 5.1 shows computational errors and CPU time for the SBM and the BEM with varied numbers of nodes (or elements), where the number of nodes in the SBM is equal to the number of elements in the BEM. The error is computed as [1]

$$L_\infty = \max_{1 \le i \le M} \left| u_i^{\text{exa}} - u_i^{\text{num}} \right|, \tag{5.4}$$

Table 5.1: Errors and CPU time for the SBM and the BEM with different numbers of nodes (or elements) for the two-dimensional heat transfer problem of integer dimension.

N	**BEM**		**SBM**	
	L_∞	**CPU time (s)**	L_∞	**CPU time (s)**
200	2.0814×10^{-2}	0.60840	3.0918×10^{-2}	0.15600
400	3.9964×10^{-3}	2.26200	5.3899×10^{-3}	0.70200
600	1.4664×10^{-3}	5.44440	1.9052×10^{-3}	2.44920
800	7.1454×10^{-4}	12.1681	9.0748×10^{-4}	6.89520
1000	4.0717×10^{-4}	19.9213	5.0949×10^{-4}	11.1853

where u_i^{exa} and u_i^{num} denote analytical and numerical solutions, respectively, at the ith test point; M is the total number of test points where both the numerical and exact solutions are evaluated.

It can be seen from Table 5.1 that these two methods have a good convergence and achieve almost the same error. On the other hand, the BEM requires twice as much CPU time as the SBM because of expensive numerical integration.

To investigate the accuracy of the SBM, we first consider the heat transfer on an integer-order dimension case, $\beta = 1$, whose exact solution is known, i. e., the solution equation (5.3). To solve the problem numerically, 600 evenly distributed source points are chosen on the boundary. Figure 5.3 displays the distributions of the analytical solution and the relative error of the temperature in the computational domain, where the surfaces are covered by 20×20 calculation points uniformly spaced over the square $\{(x, y) \mid 1 < x, y < 3\}$. We see from Figure 5.3 that the SBM can provide accurate numerical solutions.

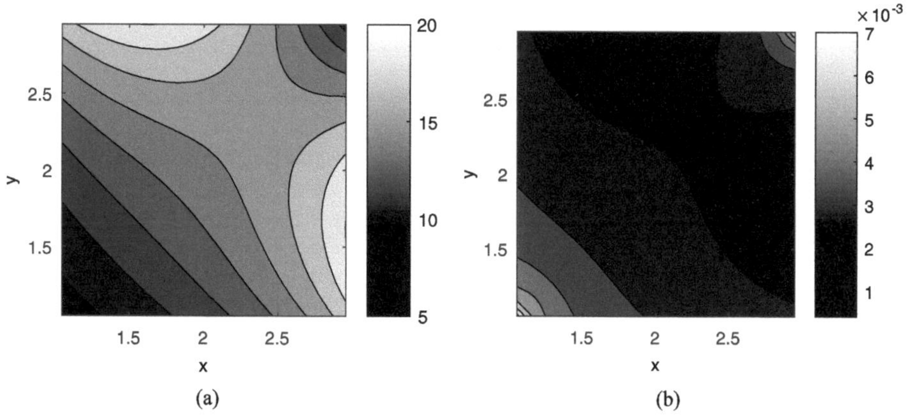

Figure 5.3: Profiles of (a) analytical solutions for the temperature and (b) relative errors of the numerical temperature by using the SBM.

To analyze the convergence of the numerical method, Figure 5.4 depicts the root mean squared relative errors (RMSREs) of SBM results with respect to the number of nodes at the abovementioned 400 calculation points. As shown in Figure 5.4, the SBM shows a good convergence with respect to an increasing number of source points. We can conclude that the SBM is accurate and convergent in the tested classical two-dimensional heat transfer problem.

Next, based on the above numerical experiment and discussions, we consider the Hausdorff Laplacian model, i. e., $\beta \neq 1$. In this case, no analytical solution is available.

Figure 5.5 displays the SBM solutions on the characteristic line $\{(x, y) \mid x = 2, 1 \leq y \leq 3\}$ in different cases with varied fractals. The SBM employs 600 evenly

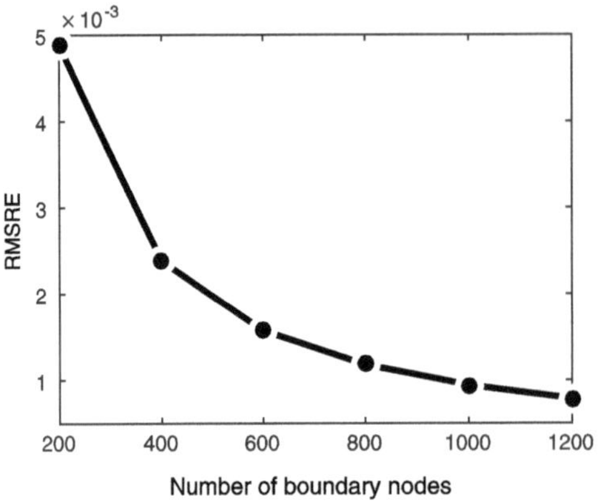

Figure 5.4: Convergence curve of the computed temperature using the SBM.

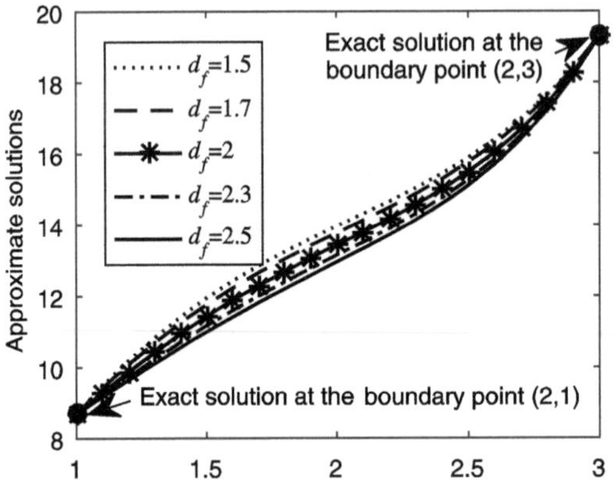

Figure 5.5: SBM solutions of temperature on the line $\{(x,y) \mid x = 2, 1 \le y \le 3\}$, where $d_f = d \times \beta = 2\beta$ represents the fractal dimensionality in the case of two topological dimensions.

distributed source points on the boundary. Note that $d_f = d \times \beta$ in Figure 5.5 represents the fractal dimensionality, where d is the topological dimension. It is observed that the SBM curves of the temperature distribution in a variety of fractal media ($d_f = 1.5, 1.7, 2, 2.3, 2.5$) gradually approach that in the standard integer-order two-dimensional media. This shows the stably continuous variation of the solutions versus the Hausdorff fractal of media.

Furthermore, the two endpoints of all curves in Figure 5.5, which are highlighted by the two small circles, represent the exact boundary conditions at the boundary points (2, 1) and (2, 3) in Figure 5.2, respectively. As we can see, all numerical solutions at the boundary points coincide with the exact boundary temperature. This indicates that the accuracy of the SBM is high and the proposed Hausdorff fractal distance fundamental solution is valid.

Example 5.2. In this case, we test a topological three-dimensional heat transfer problem in a cylinder $\{(x,y,z) \mid (x-2)^2 + (y-2)^2 \le 1, 2 \le z \le 8\}$ as shown in Figure 5.6.

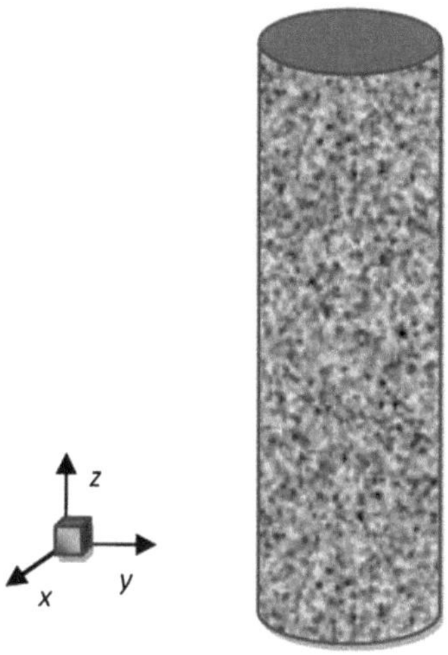

Figure 5.6: Steady heat transfer in a cylindrical fractal medium.

The temperature on the cylinder surface is assumed known and is determined by the function

$$u(x,y,z) = e^{x}\cos\left(\frac{\sqrt{2}}{2}y\right)\cos\left(\frac{\sqrt{2}}{2}z\right) + 5. \tag{5.5}$$

For the classical integer dimension, Table 5.2 lists computational errors and CPU time by using the SBM and the BEM with different numbers of nodes (or elements), in which the number of nodes in the SBM is equal to the number of elements in the BEM. As we can see in Table 5.2, these two methods give the same level of errors, but the CPU time of the SBM is significantly less than that of the BEM, as in the two-dimensional case. This clearly illustrates the computational efficiency of the SBM.

Next, we compare the SBM solution with the exact solution equation (5.5) in order to further examine the accuracy and convergence of the SBM in the calculation of three-dimensional heat transfer problems. A cross-section testing plane, $\{(x,y,z) \mid x \in$

Table 5.2: Errors and CPU time by using the SBM and the BEM with different numbers of nodes (or elements) for the three-dimensional heat transfer problem of integer dimension.

N	BEM		SBM	
	L_∞	CPU time (s)	L_∞	CPU time (s)
432	1.5739×10^{-2}	90.60538	2.3824×10^{-2}	1.170007
768	5.8217×10^{-3}	257.9477	6.5918×10^{-3}	8.408454
1200	2.8823×10^{-3}	576.3925	3.1389×10^{-3}	26.31737
1728	1.6309×10^{-3}	1196.684	2.5544×10^{-3}	78.60890
2352	1.0427×10^{-3}	2178.694	2.8396×10^{-3}	189.7128
3072	6.8039×10^{-4}	4347.639	8.9161×10^{-4}	741.4727

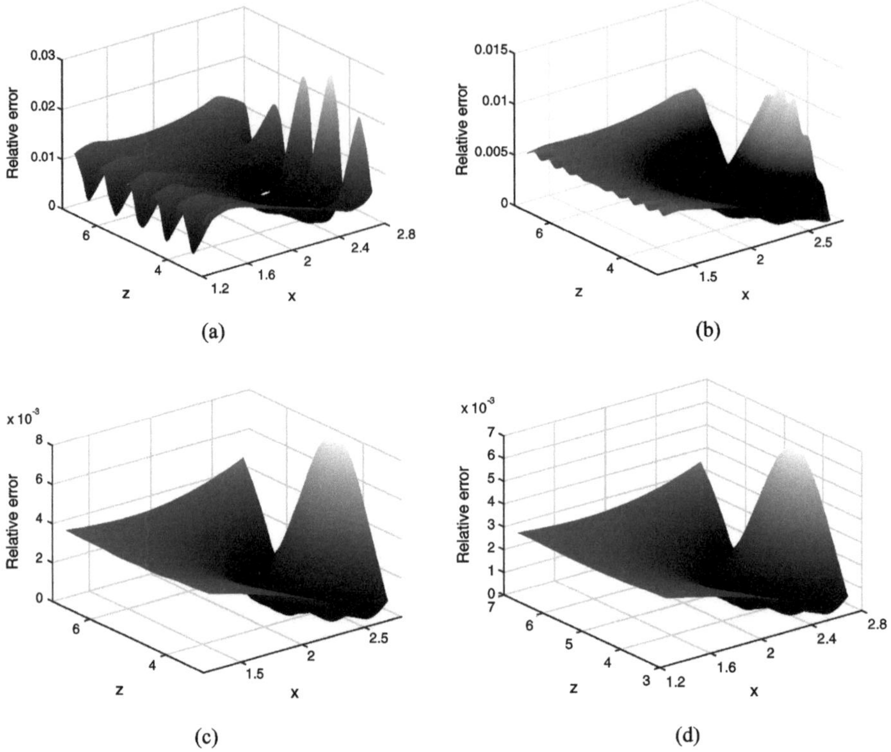

Figure 5.7: Relative error surfaces of the SBM solution on the rectangular cross-section domain $\{(x, y, z) \mid x \in [1.2, 2.8], y = 2, z \in [3, 7]\}$ within the cylinder, where the SBM boundary nodes are (a) 192, (b) 768, (c) 1728, and (d) 3072.

$[1.2, 2.8], y = 2, z \in [3, 7]\}$, within the cylinder is considered. Figure 5.7 shows the relative error surfaces of the SBM solution on the plane by using 192, 768, 1728, and 3072 boundary nodes, respectively. The error surfaces were obtained at 10×20 calculation points uniformly spaced over the square cross-section plane.

It can be observed from Figure 5.7 that with increasing source points, the relative error gradually decreases. This indicates the convergence of the SBM. Furthermore, Figure 5.8 displays a comparison of the analytical solutions and SBM results on the characteristic line $\{(x, y, z) \mid x = 2, y = 2, z \in [2, 8]\}$. It can be seen that the numerical results are in good agreement with the analytical solutions, in which the maximum relative error is 3.5×10^{-3}. This shows the accuracy of the SBM is high.

For the Hausdorff fractal media $\beta \neq 1$, the exact solution within the cylinder is unobtainable. Figure 5.9 illustrates the SBM solution on the abovementioned characteristic line $\{(x, y, z) \mid x = 2, y = 2, z \in [2, 8]\}$. It is observed from Figure 5.9 that as the fractal dimensionality $d_f = 3\beta$ increases, the solutions nearby the center of the line

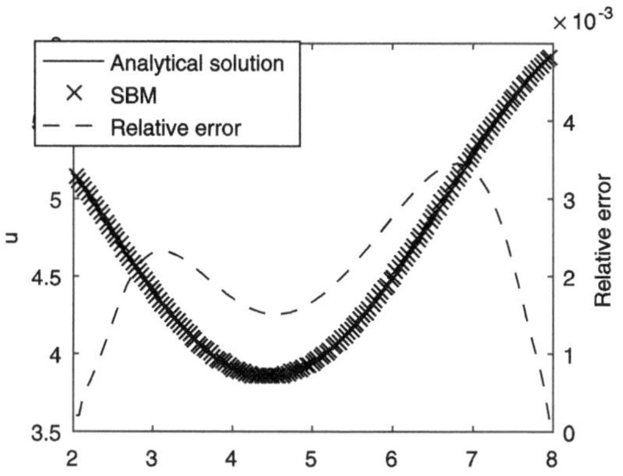

Figure 5.8: Comparison of the SBM and analytical solutions on the line $\{(x, y, z) \mid x = 2, y = 2, z \in [2, 8]\}$.

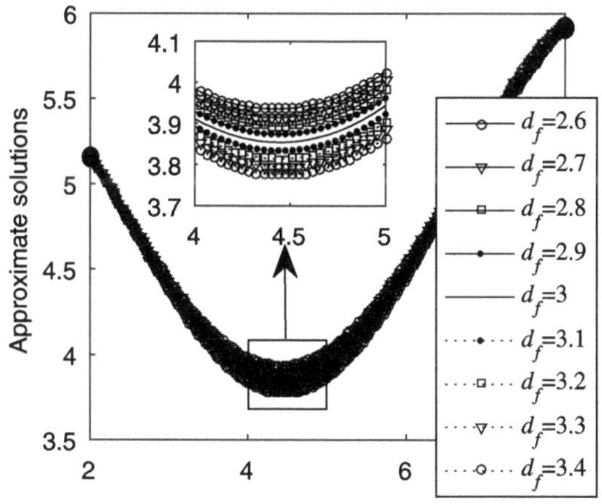

Figure 5.9: SBM solutions of temperature on the line $\{(x, y, z) \mid x = 2, y = 2, z \in [2, 8]\}$.

tend to be smaller. Figure 5.9 also clearly shows the continuous variation of the SBM solutions against varied fractal dimensionality.

The fractional Laplacian is currently a popular methodology in modeling anomalous diffusion behaviors in complex media. But its underlying relationship with the fractal remains not fully understood [36, 150]. To compare the fractal model with the fractional model, Figure 5.10 gives the SBM solutions of temperature on the same characteristic line by using the fractional Laplacian model, where s denotes the order of the fractional Laplacian. Comparing Figures 5.9 and 5.10, it can be observed that the fractional and the Hausdorff Laplacian models yield different temperature distributions. The fractional Laplacian model appears to have more temperature variation than the Hausdorff Laplacian model. It is meaningful to further investigate the relationship and difference between these models. Further work along this line requires the detailed study of experimental data to characterize the respective applicability of these two methodologies. This research is only the first step to propose a novel Hausdorff fractal distance fundamental solution model and to test its numerical properties.

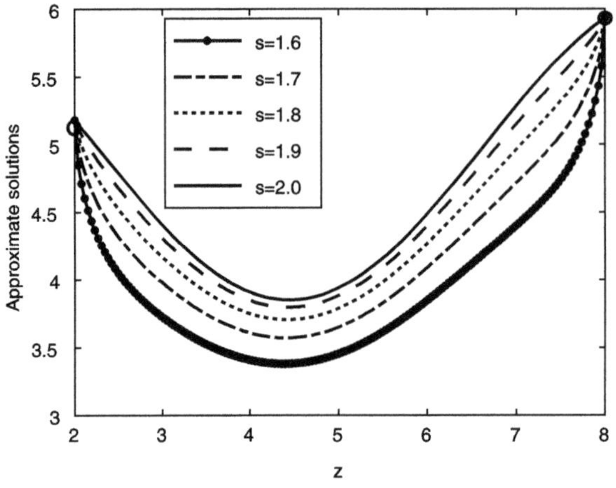

Figure 5.10: SBM solutions of temperature on the line $\{(x, y, z) \mid x = 2, y = 2, z \in [2, 8]\}$ by using the fractional Laplacian [44].

5.2 Radial basis functions

Some of the commonly used RBFs [38, 187] based on the Hausdorff fractal distance are given in Table 5.3, in which c or m represents the shape parameter, and MQ [50, 99, 100] is one of the most frequently used RBFs. Taking MQ, PS, and GA as the examples, the general Hausdorff derivatives of RBFs based on the Hausdorff fractal distance will be investigated below.

Table 5.3: List of commonly used RBFs.

RBF	Formulation
Multiquadric (MQ)	$\phi^{MQ}(r_j) = \sqrt{r_j^2 + c^2}, c > 0$
Gaussian (GA)	$\phi^{GA}(r_j) = e^{-cr_j^2}, c > 0$
Polyharmonic splines (PS) of order m	$\phi^{PS}(r_j) = r_j^{2m} \ln(r_j), m \in \mathbb{N}$
	$\phi^{PS}(r_j) = r_j^{2m-1}, m \in \mathbb{N}$

5.2.1 Multiquadric (MQ)

First we consider MQ, i. e., $\phi^{MQ}(r) = \sqrt{r^2 + c^2}$,

$$\frac{\partial \phi^{MQ}}{\partial x^\beta} = \frac{1}{\beta x^{\beta-1}} \frac{\partial \phi^{MQ}}{\partial x} = \frac{x^\beta - x_j^\beta}{\sqrt{r_j^2 + c^2}}, \tag{5.6}$$

$$\frac{\partial \phi^{\text{MQ}}}{\partial y^\beta} = \frac{1}{\beta y^{\beta-1}} \frac{\partial \phi^{\text{MQ}}}{\partial y} = \frac{y^\beta - y_j^\beta}{\sqrt{r_j^2 + c^2}}, \tag{5.7}$$

$$\frac{\partial}{\partial x^\beta}\left(\frac{\partial \phi^{\text{MQ}}}{\partial x^\beta}\right) = \frac{1}{\beta x^{\beta-1}} \frac{\partial(\frac{x^\beta - x_j^\beta}{\sqrt{r^2+c^2}})}{\partial x} = \frac{(x^\beta - x_j^\beta)^2 + c^2}{(r_j^2 + c^2)^{3/2}}, \tag{5.8}$$

$$\frac{\partial}{\partial y^\beta}\left(\frac{\partial \phi^{\text{MQ}}}{\partial y^\beta}\right) = \frac{1}{\beta y^{\beta-1}} \frac{\partial(\frac{y^\beta - y_j^\beta}{\sqrt{r^2+c^2}})}{\partial y} = \frac{(y^\beta - y_j^\beta)^2 + c^2}{(r_j^2 + c^2)^{3/2}}, \tag{5.9}$$

$$\frac{\partial}{\partial x^\beta}\left(\frac{\partial \phi^{\text{MQ}}}{\partial x^\beta}\right) + \frac{\partial}{\partial y^\beta}\left(\frac{\partial \phi^{\text{MQ}}}{\partial y^\beta}\right) = \frac{(x^\beta - x_j^\beta)^2 + c^2}{(r_j^2 + c^2)^{3/2}} + \frac{(y^\beta - y_j^\beta)^2 + c^2}{(r_j^2 + c^2)^{3/2}} = \frac{r_j^2 + c^2}{(r_j^2 + c^2)^{3/2}}. \tag{5.10}$$

5.2.2 Polyharmonic splines (PS)

Next we consider PS, i. e., $\phi^{\text{PS}}(r_j) = r^{2m} \ln(r_j)$, $m \in \mathbb{N}$,

$$\frac{\partial \phi^{\text{PS}}}{\partial x^\beta} = \frac{1}{\beta x^{\beta-1}} \frac{\partial \phi^{\text{PS}}}{\partial x} = m(x^\beta - x_j^\beta) r_j^{2(m-1)} \ln(r_j^2 + 1), \tag{5.11}$$

$$\frac{\partial \phi^{\text{PS}}}{\partial y^\beta} = \frac{1}{\beta y^{\beta-1}} \frac{\partial \phi^{\text{PS}}}{\partial y} = m(y^\beta - y_j^\beta) r_j^{2(m-1)} \ln(r_j^2 + 1), \tag{5.12}$$

$$\begin{aligned}\frac{\partial}{\partial x^\beta}\left(\frac{\partial \phi^{\text{PS}}}{\partial x^\beta}\right) &= \frac{1}{\omega x^{\beta-1}} \frac{\partial(m(x^\beta - x_j^\beta) r_j^{2(m-1)} \ln(r_j^2 + 1))}{\partial x} \\ &= r^{2(m-2)}(2m(m \ln r_j^2 + 2)(x^\beta - x_j^\beta)^2 + (m \ln r_j^2 + 1)((y^\beta - y_j^\beta)^2 - (x^\beta - x_j^\beta)^2)),\end{aligned} \tag{5.13}$$

$$\begin{aligned}\frac{\partial}{\partial y^\phi}\left(\frac{\partial \phi^{\text{PS}}}{\partial y^\beta}\right) &= \frac{1}{\beta y^{\beta-1}} \frac{\partial(m(y^\beta - y_j^\beta) r_j^{2(\beta-1)} \ln(r_j^2 + 1))}{\partial y} \\ &= r^{2(m-2)}(2m(m \ln r_j^2 + 2)(y^\beta - y_j^\beta)^2 + (m \ln r_j^2 + 1)((x^\beta - x_j^\beta)^2 - (y^\beta - y_j^\beta)^2)),\end{aligned} \tag{5.14}$$

$$\frac{\partial}{\partial x^\beta}\left(\frac{\partial \phi^{\text{PS}}}{\partial x^\beta}\right) + \frac{\partial}{\partial y^\beta}\left(\frac{\partial \phi^{\text{PS}}}{\partial y^\beta}\right) = r_j^{2(m-1)} 2m(m \ln r_j^2 + 2). \tag{5.15}$$

5.2.3 Inverse multiquadric (IMQ) and Gaussian (GA)

Similarly, we can derive the Hausdorff derivatives of IMQ and GA RBF, i. e.,

$$\frac{\partial}{\partial x^\beta}\left(\frac{\partial \phi^{\text{IMQ}}}{\partial x^\beta}\right) + \frac{\partial}{\partial y^\beta}\left(\frac{\partial \phi^{\text{IMQ}}}{\partial y^\beta}\right) = (r_j^2 - 2c^2)/(r_j^2 + c^2)^{5/2}, \tag{5.16}$$

$$\frac{\partial}{\partial x^\beta}\left(\frac{\partial \phi^{\text{GA}}}{\partial x^\beta}\right) + \frac{\partial}{\partial y^\beta}\left(\frac{\partial \phi^{\text{GA}}}{\partial y^\beta}\right) = 4ce^{-cr_j^2}(cr_j^2 - 1). \tag{5.17}$$

5.3 The Kansa method on the Hausdorff fractal distance

With the help of the Hausdorff fractal distance, the Kansa method is very easy to implement in solving Hausdorff partial differential equations, such as Hausdorff Laplacian and Hausdorff Poisson problems [187]. Consider the following two-dimensional Hausdorff Poisson equations in a bounded domain $\Omega \subset \mathbb{R}^2$:

$$\frac{\partial}{\partial x^\beta}\left(\frac{\partial u(x,y)}{\partial x^\beta}\right)+\frac{\partial}{\partial y^\beta}\left(\frac{\partial u(x,y)}{\partial y^\beta}\right)=f(x,y), \quad (x,y)\in\Omega, \tag{5.18}$$

$$u(x,y)=g(x,y), \quad (x,y)\in\Gamma. \tag{5.19}$$

According to the Kansa method, we collocate $N = Ni + Nb$ different points $\{(x_j,y_j)\}_{j=1}^N$ in $\bar{\Omega}$, of which $\{(x_j,y_j)\}_{j=1}^{Ni}$ are interior points and $\{(x_j,y_j)\}_{j=Ni+1}^N$ are boundary points. The approximate expansion of $u(x,y)$ is

$$u(x,y)=\sum_{j=1}^{N}\lambda_j\phi(r_j), \tag{5.20}$$

where λ_j are unknown expansion coefficients to be evaluated, $\phi(r_j)$ is the RBF, and $r_j=\sqrt{(x^\beta-x_j^\beta)^2+(y^\beta-y_j^\beta)^2}$ represents the Hausdorff fractal distance.

By substituting equation (5.20) into equations (5.18)–(5.19), we have

$$\sum_{j=1}^{N}\lambda_j\left(\frac{\partial}{\partial x^\beta}\left(\frac{\partial\phi}{\partial x^\beta}\right)+\frac{\partial}{\partial y^\beta}\left(\frac{\partial\phi}{\partial y^\beta}\right)\right)=f(x_i,y_i), \quad i=1,2,\dots,Ni, \tag{5.21}$$

$$\sum_{j=1}^{N}\lambda_j\phi(r_j)=g(x_i,y_i), \quad i=Ni+1,\dots,N, \tag{5.22}$$

from which we can solve the $N\times N$ linear system for the unknowns $\{\lambda_j\}_{j=1}^N$. Then equation (5.20) can give us the approximate solution at any point in the domain.

5.3.1 Shape parameters

The shape parameters of RBFs can have a significant influence on the accuracy and robustness of the solution when applying the RBFs, and the method for determining values of the shape parameter is needed. Because of this, several parameter optimization approaches have been developed [63, 80, 157]. Hardy [80] recommended the value $c = 0.815d$ as the optimal parameter of the MQ, where d is the average distance between nodes calculated by $(1/N)\sum_{i=1}^N d_i$, where d_i denotes the distance from the node $\mathbf{x}_i$ to its nearest neighbor. Fasshauer [61] suggested the formulation for the MQ RBF shape parameters, $c = 1.25D/\sqrt{N}$, where D represents the diameter of the circle that contains all the nodes and N is the number of the nodes. Franke [65] presented the formula $c = 2/\sqrt{N}$ to determine the MQ RBF shape parameters. For the details on the shape parameter optimization approaches, readers are referred to [62].

In this work, we mainly develop the most used algorithm, known as "leave-one-out" cross-validation (LOOCV), proposed by Rippa [157], for the Kansa method based on the Hausdorff fractal distance. The LOOCV first requires choosing a known sample function. The MATLAB code "LOOCV2D" provided in [62] uses the function $f_s = \operatorname{sinc}(x)\operatorname{sinc}(y)$ as the test function. Based on the Hausdorff derivative, this study defines the following test function:

$$f_s(x,y) = \operatorname{sinc}(x^\beta)\operatorname{sinc}(y^\beta), \tag{5.23}$$

where β is the order of the fractal derivative.

Then we define the following vector of data sites without the point $\mathbf{x}_k$:

$$\mathbf{x}^{[k]} = [\mathbf{x}_1, \ldots, \mathbf{x}_{k-1}, \mathbf{x}_{k+1}, \ldots \mathbf{x}_N]^T. \tag{5.24}$$

Similarly, we define $\mathbf{f}^{[k]}$, $P_f^{[k]}$, and $\boldsymbol{\lambda}^{[k]}$ in the LOOCV algorithm later on.

Let $P_f^{[k]}$ be the partial RBF interpolant to the data $\mathbf{f}^{[k]}$, i. e.,

$$P_f^{[k]}(\mathbf{x}) = \sum_{j=1}^{N-1} \lambda_j^{[k]} \phi(r_j(\mathbf{x}, \mathbf{x}_j^{[k]})), \tag{5.25}$$

where $r_j(\mathbf{x}, \mathbf{x}_j^{[k]})$ is the Hausdorff fractal distance, and if E_k is the error

$$E_k = f_s(\mathbf{x}_k) - P_f^{[k]}(\mathbf{x}_k), \tag{5.26}$$

then the quality of the overall fit to the entire data set will be determined by the norm of the vector of errors $E = [E_1, \ldots, E_N]^T$ obtained by removing in turn each one of the data points and comparing the resulting fit with the (known) value at the removed point as described above. In [157], the author presented several examples where he uses the l_1 and l_2 norms. In our case, we will use the maximum norm.

By adding a loop over c (for MQ, IMQ, and GA) or m (for PS) we can compare the error norms for different values of the shape parameter and choose that value of c or m that yields the minimal error norm as the optimal one. However, it is well known that the implementation of the above procedure may be expensive. In view of computational complexity, the computation of the error components can be simplified to a single formula, i. e.,

$$E_k = \frac{\lambda_k}{A_{kk}^{-1}}, \tag{5.27}$$

where λ_k is the kth coefficient in the expansion of the interpolant P_f based on the full data set and A_{kk}^{-1} is the kth diagonal element of the inverse of the corresponding interpolation matrix.

5.3.2 Numerical experiments

In this section, four benchmark numerical examples on the Hausdorff derivative Poisson equations are examined to verify the methodology developed above. The first two examples compare different algorithms for determining shape parameters of RBFs and find an efficient and accurate scheme to choose the shape parameters of RBFs in the Kansa method based on the Hausdorff fractal distance. Using irregular node distribution, the third example investigates the Hausdorff derivative Poisson equations in the irregular domain. The last example examines the multiconnected domain problems.

Example 5.3. First, we consider the following Hausdorff derivative Poisson equation in a unit square $\Omega = [0,1] \times [0,1]$:

$$\frac{\partial}{\partial x^{0.8}}\left(\frac{\partial u(x,y)}{\partial x^{0.8}}\right) + \frac{\partial}{\partial y^{0.8}}\left(\frac{\partial u(x,y)}{\partial y^{0.8}}\right) = 6x^{0.8} + 2, \quad (x,y) \in \Omega, \tag{5.28}$$

$$\begin{cases} u(0,y) = y^{1.6}, \\ u(1,y) = 1 + y^{1.6}, \\ u(x,0) = x^{2.4}, \\ u(x,1) = x^{2.4} + 1. \end{cases} \tag{5.29}$$

Its exact solution is $u(x,y) = x^{2.4} + y^{1.6}$. Selection of appropriate shape parameters of RBFs is extremely important to ensure accuracy while solving equations using the Kansa method. In order to compare the accuracy of the MQ Kansa method using different approaches for calculating the shape parameter c, Figure 5.11 shows the maximum absolute errors of solution at 20×20 test points uniformly distributed over the square

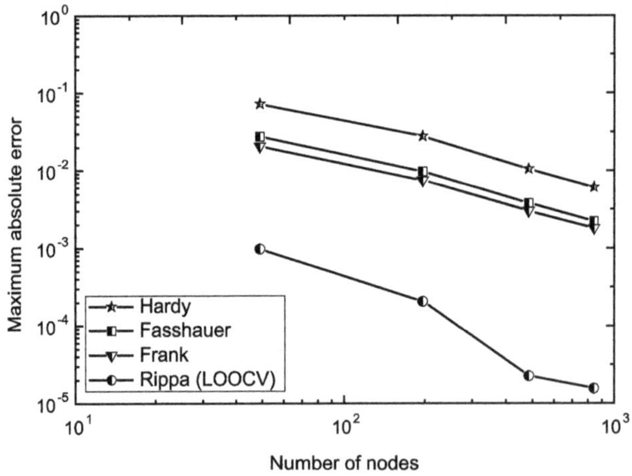

Figure 5.11: Maximum absolute errors of MQ Kansa solutions with different shape parameter optimization.

$[0,1] \times [0,1]$ obtained by using equispaced nodes and the MQ Kansa method with different parameter optimization. It is observed that all four approaches are efficient for the determination of the shape parameter and yield convergent numerical solutions when increasing the number of nodes. Furthermore, it can be clearly seen that the LOOCV is more accurate than the other three approaches, and the error declines $1 \sim 2$ orders of magnitude.

Next, we compare the accuracy of Kansa's method using different RBFs, in which the optimal shape parameters are determined by the LOOCV approach. Figure 5.12 displays the maximum absolute errors of numerical solutions obtained by using the Kansa method with different RBFs. The optimal shape parameters obtained by the LOOCV are listed in Table 5.4. As can be seen from Figure 5.12, the MQ RBF achieves higher accuracy and better convergence than the other RBFs. From Table 5.4, it can be seen that the optimal shape parameters gradually decline with increasing numbers of nodes for the MQ, IMQ, and GA. On the contrary, the optimal shape parameters of the PS increase as the number of nodes increases.

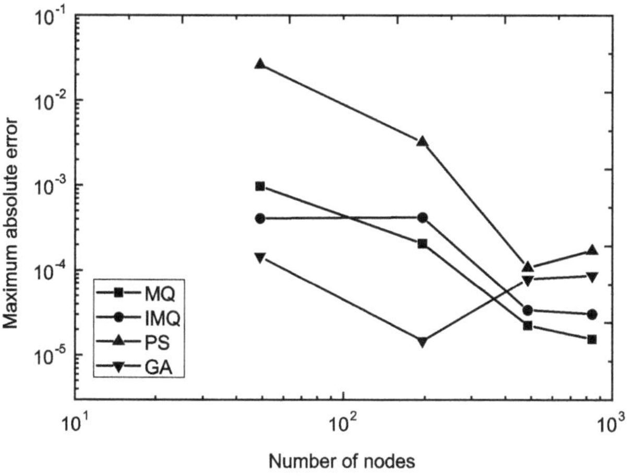

Figure 5.12: Maximum absolute errors of the Kansa solutions with different RBFs.

Table 5.4: Optimal shape parameters of RBFs obtained by using the LOOCV.

N	MQ	IMQ	PS	GA
49	2.219950	1.315879	2	1.014523
196	0.813618	1.014523	3	0.914070
484	0.612714	0.914070	5	0.311357
841	0.411809	0.330481	5	0.210905

Example 5.4. Next, we consider

$$\frac{\partial}{\partial x^{\beta}}\left(\frac{\partial u(x,y)}{\partial x^{\beta}}\right)+\frac{\partial}{\partial y^{\beta}}\left(\frac{\partial u(x,y)}{\partial y^{\beta}}\right)=4, \quad (x,y)\in[1,2]^2, \tag{5.30}$$

$$\begin{cases} u(1,y) = 1 + y^{2\beta}, \\ u(2,y) = 4^{\beta} + y^{2\beta}, \\ u(x,1) = x^{2\beta} + 1, \\ u(x,2) = x^{2\beta} + 4^{\beta}, \end{cases} \tag{5.31}$$

where β is the order of the fractal derivative. The exact solution is $u(x,y) = x^{2\beta} + y^{2\beta}$.

To investigate the influence of the node distribution on the numerical solution, we consider the two types of node distribution shown in Figure 5.13, i. e., regular nodes and irregular nodes. Here the irregular nodes are derived by jiggling the regular nodes, i. e., assigning a perturbation on regular nodes along x- and y-directions. Table 5.5 displays the maximum absolute errors of the Kansa solutions with different RBFs and orders of the fractal derivative, where $N = 196$. The optimal shape parameters determined by the LOOCV approach are shown in Table 5.6. We can see in Table 5.5 that the four RBFs have higher accuracy for different types of node distribution and orders of the fractal derivative. Compared with the other three RBFs, on the whole the MQ RBF is more accurate and stable for various orders of the fractal derivative, for regular and irregular nodes. Furthermore, it can be seen in Table 5.6 that the optimal shape parameters estimated by the LOOCV increase with increasing orders of the fractal derivative for the MQ, IMQ, and GA.

By applying the irregular nodes and MQ RBF, the numerical results are compared with the analytical solution on the computational domain. Figure 5.14 displays the profiles of the analytical solutions and the numerical solutions for the various orders of the fractal derivative. As can be seen, the numerical results agree quite well with

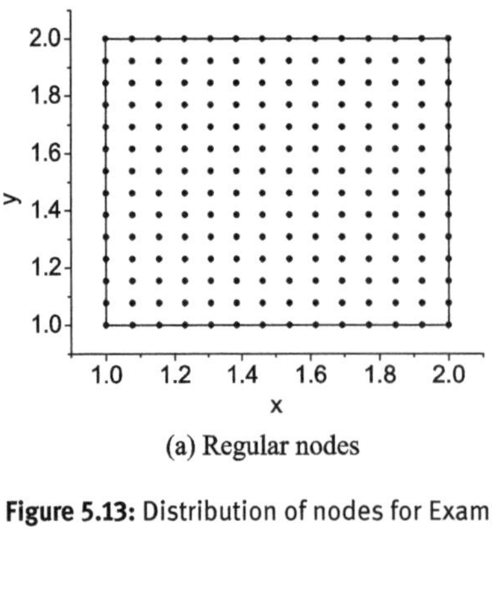

(a) Regular nodes

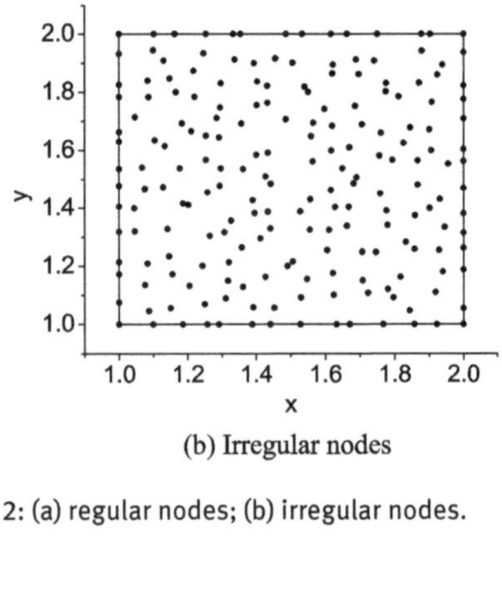

(b) Irregular nodes

Figure 5.13: Distribution of nodes for Example 2: (a) regular nodes; (b) irregular nodes.

Table 5.5: Maximum absolute errors of the Kansa solutions with different RBFs and orders of the fractal derivative.

β	Regular nodes				Irregular nodes			
	MQ	IMQ	PS	GA	MQ	IMQ	PS	GA
0.1	1.4e−5	4.2e−5	2.5e−4	2.7e−4	3.0e−5	1.3e−4	1.3e−4	3.1e−4
0.2	2.9e−6	6.4e−6	3.3e−4	1.4e−4	1.1e−5	3.2e−5	1.7e−4	2.6e−4
0.3	1.8e−6	2.3e−6	7.1e−4	1.0e−4	1.6e−5	1.5e−5	4.9e−4	3.3e−4
0.4	1.1e−5	8.9e−6	1.3e−4	2.9e−5	2.5e−6	2.9e−5	2.5e−4	1.7e−4
0.5	2.6e−5	1.1e−5	4.9e−3	1.7e−5	6.3e−6	4.4e−5	2.2e−4	2.3e−5
0.6	2.2e−6	8.8e−5	8.3e−4	3.3e−5	9.2e−7	1.1e−4	6.6e−4	3.0e−5
0.7	8.6e−5	1.5e−4	7.4e−4	2.9e−5	7.4e−5	1.2e−4	1.8e−3	4.3e−5
0.8	6.2e−5	1.4e−4	2.5e−3	2.9e−5	2.2e−5	1.7e−4	3.3e−3	6.0e−5
0.9	1.1e−5	1.9e−4	4.6e−4	3.0e−5	7.0e−6	5.1e−5	4.3e−3	2.3e−5
1	8.2e−5	1.7e−4	3.4e−4	5.1e−6	1.5e−4	1.1e−4	1.6e−3	1.5e−6

Table 5.6: Optimal shape parameters of the RBFs obtained by the LOOCV.

β	Regular nodes				Irregular nodes			
	MQ	IMQ	PS	GA	MQ	IMQ	PS	GA
0.1	0.110452	0.110452	4	0.110452	0.110452	0.110452	4	0.110452
0.2	0.210904	0.210905	4	0.110452	0.210905	0.210905	4	0.110452
0.3	0.311357	0.311357	4	0.210905	0.311357	0.311357	4	0.110452
0.4	0.411809	0.411809	4	0.210905	0.411809	0.512261	4	0.210905
0.5	0.512261	0.512261	6	0.311357	0.512261	0.512261	4	0.311357
0.6	0.713166	0.813618	5	0.411809	0.512261	0.713166	6	0.411809
0.7	0.813619	1.014523	4	0.411809	0.813618	1.014523	4	0.411809
0.8	0.813619	1.516784	3	0.512714	0.914070	1.114975	3	0.612714
0.9	0.914070	1.617236	4	0.612261	1.014523	1.617236	3	0.713166
1	1.114974	1.617236	4	0.914070	1.315879	1.617236	4	0.713166

the corresponding analytical solutions, where the maximum absolute errors for $\beta = 0.1$ and $\beta = 0.9$ are 3.01×10^{-5} and 6.95×10^{-6}, respectively. Thus, we can carefully conclude that the MQ Kansa method can be considered as a powerful and accurate tool for the calculation of Hausdorff derivative Poisson equations.

Example 5.5. In this example we consider the following Poisson equation:

$$\frac{\partial}{\partial x^\beta}\left(\frac{\partial u(x,y)}{\partial x^\beta}\right) + \frac{\partial}{\partial y^\beta}\left(\frac{\partial u(x,y)}{\partial y^\beta}\right) = 6x^\beta - y^\beta + 10, \quad (x,y) \in \Omega, \tag{5.32}$$

$$u(x,y) = e^{x^\beta} \sin y^\beta + x^{3\beta} - y^{3\beta}/6 + 5y^{2\beta}, \quad (x,y) \in \Gamma, \tag{5.33}$$

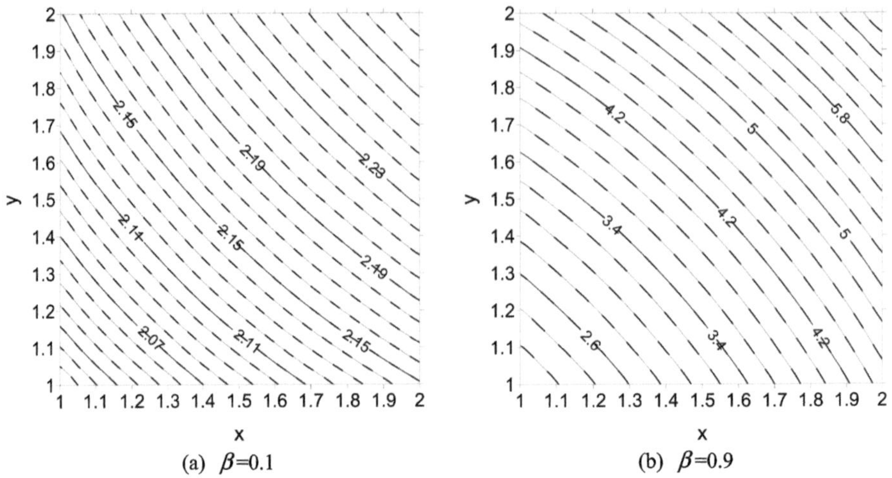

Figure 5.14: Comparison of exact and numerical results with irregular nodes and $N = 196$, where the solid line (—) represents exact solutions and the dotted line (– –) represents Kansa solutions.

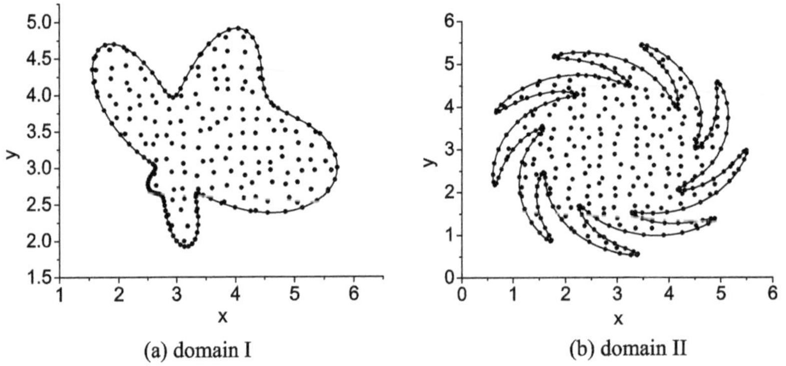

Figure 5.15: Geometry of the problem and the configuration of the nodes distribution: (a) amoeba-like domain; (b) gear-shape domain.

where β is the order of the fractal derivative and Ω is an irregular shape domain with the boundary Γ shown in Figure 5.15. The parametric equations of the irregular boundaries Γ are defined as follows:

$$\Gamma = \{(x, y) \mid x = 3 + \rho(\theta)\cos\varphi(\theta), y = 3 + \rho(\theta)\sin\varphi(\theta), 0 \le \theta \le 2\pi\}, \tag{5.34}$$

where

$$\rho(\theta) = e^{\sin\theta}\sin^2(2\theta) + e^{\cos\theta}\cos^2(2\theta), \quad \varphi(\theta) = \theta, \tag{5.35}$$

is the amoeba-like domain (domain I) and

$$\rho(\theta) = 2 + \frac{1}{2}\sin(9\theta), \quad \varphi(\theta) = \theta + \frac{1}{2}\sin(9\theta), \tag{5.36}$$

is the gear-shaped domain (domain II).

The exact solution is $u(x,y) = e^{x^\beta}\sin y^\beta + x^{3\beta} - y^{3\beta}/6 + 5y^{2\beta}$. Using the MQ Kansa method with the irregular nodes shown in Figure 5.15, Figure 5.16 presents the comparisons of numerical and analytical results for the various orders of the fractal derivative, in which the results along the circle $\{(x,y) \mid x = 3.5 + 0.5\cos\theta,\ y = 3.5 + 0.5\sin\theta, 0 \le \theta \le 2\pi\}$ for domain I are given in Figure 5.15(a); the results along the circle $\{(x,y) \mid x = 3 + 1.2\cos\theta, y = 3 + 1.2\sin\theta, 0 \le \theta \le 2\pi\}$ for domain II are displayed in Figure 5.15(b). For the amoeba-like domain, $Ni = 143$ and $Nb = 100$ nodes are chosen on and inside the boundary, respectively; for the gear-shaped domain, $Ni = 180$ and $Nb = 200$ nodes are chosen on and inside the boundary, respectively. It can be seen from Figure 5.16 that the numerical solutions coincide with the exact solutions for various orders of the fractal derivative.

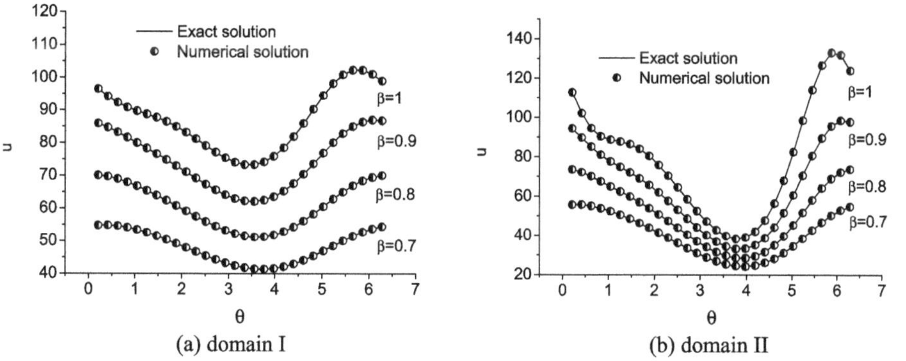

Figure 5.16: Comparisons of exact and analytical results for the various orders of the fractal derivative: (a) along the circle $\{(x,y) \mid x = 3.5 + 0.5\cos\theta,\ y = 3.5 + 0.5\sin\theta, 0 \le \theta \le 2\pi\}$ for domain I; (b) along the circle $\{(x,y) \mid x = 3 + 1.2\cos\theta, y = 3 + 1.2\sin\theta, 0 \le \theta \le 2\pi\}$ for domain II.

Figures 5.17 and 5.18 show the distributions of the exact solution and relative error in domain I and domain II under $\beta = 0.7$, respectively. From these figures we can see that the MQ Kansa method based on the Hausdorff fractal distance yields very accurate numerical results, in which the largest relative errors are less than 2.6×10^{-5} and 2.2×10^{-4} for the two considered domains, respectively.

Example 5.6. As in Example 5.5, we consider the following Poisson equation in a multiply connected domain Ω shown in Figure 5.19:

$$\frac{\partial}{\partial x^\beta}\left(\frac{\partial u(x,y)}{\partial x^\beta}\right) + \frac{\partial}{\partial y^\beta}\left(\frac{\partial u(x,y)}{\partial y^\beta}\right) = -2\sin(x^\beta) + 6y^\beta, \quad (x,y) \in \Omega, \tag{5.37}$$

$$u(x,y) = e^{y^\beta}\cos x^\beta + 2\sin(x^\beta) + y^{3\beta}, \quad (x,y) \in \Gamma, \tag{5.38}$$

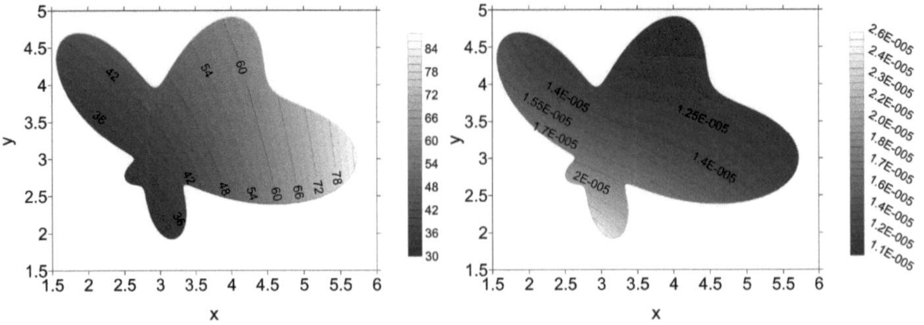

(a) exact (b) relative error

Figure 5.17: Distributions of the exact solution and relative error for the amoeba-like domain.

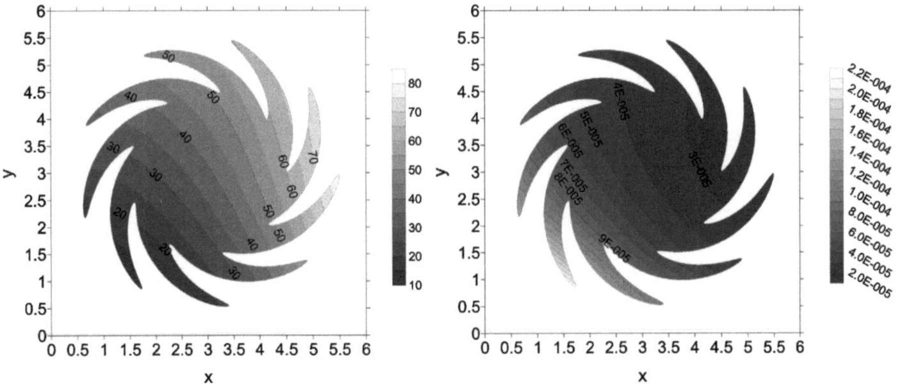

(a) exact (b) relative error

Figure 5.18: Distributions of the exact solution and relative error for the gear-shaped domain.

where β is the order of the fractal derivative and Ω is an irregular multiply connected domain with the boundary $\Gamma = \Gamma_1 + \Gamma_2 + \Gamma_3$. The parametric equations of the irregular boundaries $\Gamma_1, \Gamma_2, \Gamma_3$ are defined as follows:

$$\Gamma_1 = \{(x, y) \mid x = \rho_{11}(\theta)\varphi_{11}(\theta), y = \rho_{12}(\theta)\varphi_{12}(\theta), 0 \le \theta \le 2\pi\}, \tag{5.39}$$

$$\Gamma_2 = \{(x, y) \mid x = -3 + \rho_{21}(\theta)\varphi_{21}(\theta), y = \rho_{22}(\theta)\varphi_{22}(\theta), 0 \le \theta \le 2\pi\}, \tag{5.40}$$

$$\Gamma_3 = \{(x, y) \mid x = 3 + \rho_{31}(\theta)\varphi_{31}(\theta), y = \rho_{32}(\theta)\varphi_{32}(\theta), 0 \le \theta \le 2\pi\}, \tag{5.41}$$

where

$$\begin{aligned}
&\rho_{11}(\theta) = \rho_{12}(\theta) = 4\sqrt{\cos 2\theta + \sqrt{1.1 - \sin^2 2\theta}}, \quad \varphi_{11}(\theta) = \cos\theta, \quad \varphi_{12}(\theta) = \sin\theta, \\
&\rho_{21}(\theta) = \rho_{22}(\theta) = \left(\cos 3\theta + \sqrt{2 - \sin^2 3\theta}\right)^{1/3}, \quad \varphi_{21}(\theta) = \cos\theta, \quad \varphi_{22}(\theta) = \sin\theta, \\
&\rho_{31}(\theta) = \rho_{32}(\theta) = 0.5, \quad \varphi_{31}(\theta) = 2\cos\theta - \cos 2\theta, \quad \varphi_{32}(\theta) = 2\sin\theta - \sin 2\theta.
\end{aligned} \tag{5.42}$$

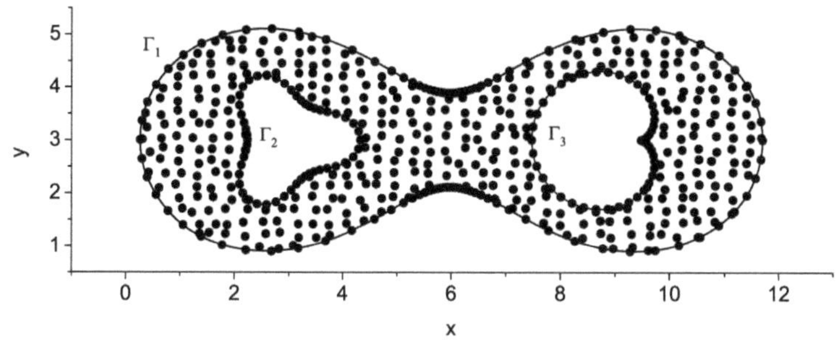

Figure 5.19: Geometry of the problem and the configuration of the nodes distribution.

Using the MQ Kansa method with the irregular nodes shown in Figure 5.19, and applying the LOOCV scheme to determine the optimal shape parameter, we examine the accuracy and convergence of the proposed method for this multiply connected domain.

First, we take $Nb = 200$ nodes, evenly distributed on the boundary according to the angle (100 nodes on Γ_1; 50 nodes on Γ_2; 50 nodes on Γ_3), and $Ni = 401$ irregular nodes in the computational domain. Figure 5.20 compares the numerical and analytical results along the circle $\{(x,y) \mid x = 3 + 1.6\cos\theta, y = 3 + 1.6\sin\theta, 0 \le \theta \le 2\pi\}$ for the various orders of the fractal derivative. As can be seen from Figure 5.20, the numerical solutions completely coincide with the exact solutions for various orders of the fractal derivative. This clearly shows that the MQ Kansa method based on the Hausdorff fractal distance, in conjunction with the LOOCV approach, is accurate for Hausdorff derivative Poisson equations in the irregular multiply connected domain.

To investigate the convergence of the MQ Kansa method based on the Hausdorff fractal distance, Figure 5.21 displays the distributions of the exact solution and relative errors obtained by using 315, 601, and 1030 nodes in the domain with $\beta = 0.9$,

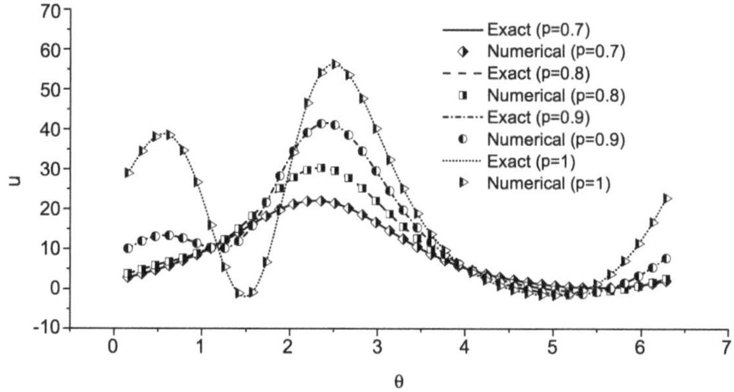

Figure 5.20: Comparison of numerical and analytical results along the circle $\{(x,y) \mid x = 3 + 1.6\cos\theta, y = 3 + 1.6\sin\theta, 0 \le \theta \le 2\pi\}$ for various orders of the Hausdorff derivative.

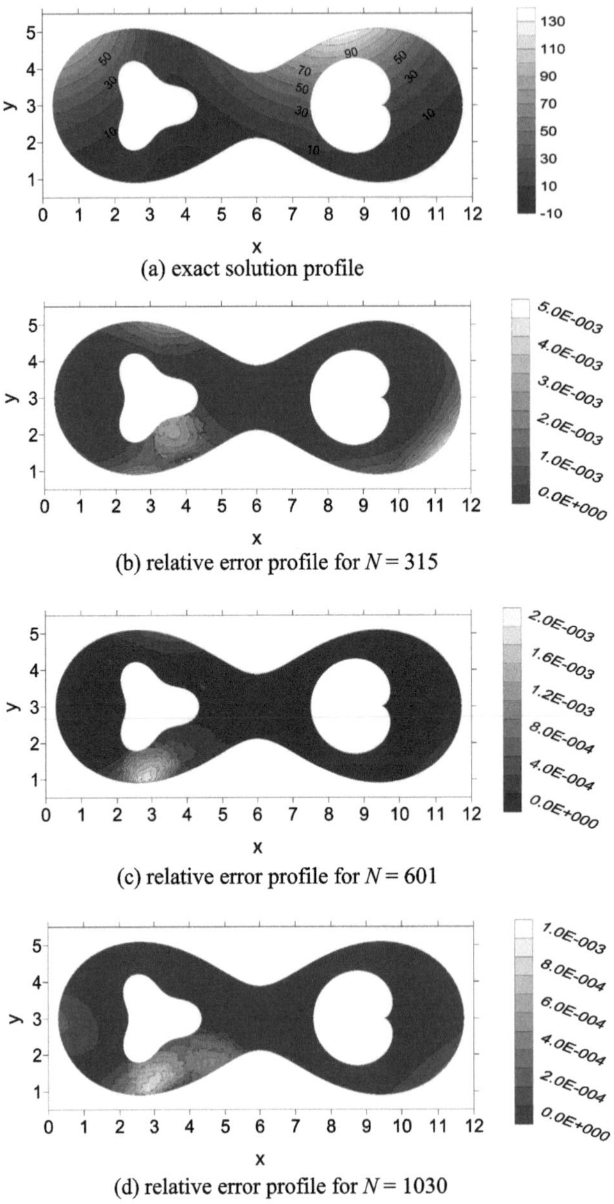

Figure 5.21: Distributions of the exact solution and relative error in the computational domain with $\omega = 0.9$: (a) exact solution; (b) relative error for $N = 315$; (c) relative error for $N = 601$; (d) relative error for $N = 1030$.

respectively. From these figures it can be observed that the MQ Kansa method based on the Hausdorff fractal distance is convergent with respect to increasing numbers of nodes. Furthermore, we can also see that the method achieves high accuracy.

6 Hausdorff PDE modeling

6.1 Anomalous diffusion

In recent years, anomalous diffusion has been widely observed, in processes such as pollutant and gas transport in porous media [9, 18]. The phenomenon is believed to behave as a non-Markovian process, induced by the heterogeneity of the transmission media. The mean square displacement for the anomalous diffusion process behaves in accordance with $\langle x^2 \rangle \propto t^\alpha$ with $\alpha \neq 1$, which is beyond the description of the well-known classical diffusion model [177]. Nevertheless, the existing Hausdorff diffusion model is limited to the one-dimensional case [45], and the exploration on physical modeling of the distance in complex materials is still ongoing. This section will propose the three-dimensional diffusion model on the basis of the Hausdorff derivative and derive the corresponding fundamental solution.

6.1.1 The Hausdorff derivative diffusion equation

As mentioned in references [31, 177], the laws of physics are invariant regardless of spatial or temporal fractal metrics, which will result in the anomalous behaviors. The anomalous diffusion process disobeys the Fick's second law and the mean square displacement is a non-linear function of t. It is generally agreed that the physical mechanism of anomalous diffusion is due to the fractal structure or fractal distribution of material components. Recently, the Hausdorff diffusion equation in one dimension has been proposed to characterize anomalous diffusion behaviors [45]. It is formulated as

$$\frac{\partial u(x,t)}{\partial t^\alpha} = D\frac{\partial}{\partial x^\beta}\left(\frac{\partial u(x,t)}{\partial x^\beta}\right), \tag{6.1}$$

where α and β represent the fractals in time and space, respectively, and D is the diffusion coefficient. For the mean square displacement for equation (6.1) it follows that $\langle x^2 \rangle \propto t^{(3\alpha-\alpha\beta)/2\beta}$. In this representation, the Hausdorff derivative orders α and β can characterize the sub-diffusion and super-diffusion, as shown in Figure 6.1. If $\alpha = \beta = 1$, the Hausdorff fractal diffusion model degenerates to the classical diffusion equation with $\langle x^2 \rangle \propto t$.

With the help of Hausdorf derivatives, equation (6.1) can be reformulated as [22]

$$\frac{1}{\alpha t^\alpha}\frac{\partial u(x,t)}{\partial t} = \frac{D}{\beta^2 x^{\beta-1}}\left[\frac{1-\beta}{x^\beta}\frac{\partial u(x,t)}{\partial x} + \frac{1}{x^{\beta-1}}\frac{\partial^2 u(x,t)}{\partial x^2}\right]. \tag{6.2}$$

It is obvious that equation (6.2) is formally a kind of convection–diffusion equation with time- and space-dependent characteristics. It is reasonable to deduce that the Hausdorff diffusion equation can characterize the anomalous diffusion behavior.

https://doi.org/10.1515/9783110608526-006

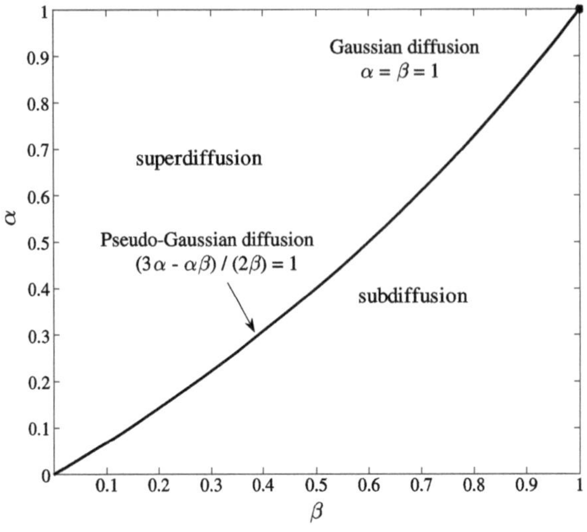

Figure 6.1: Anomalous diffusion phase diagram with the fractal derivative orders in space β and time α. The point (1,1) corresponds to Gaussian or normal diffusion, while all the points along the curve drawn between (0,0) and (1,1) represent anomalous diffusion, which we give a linear mean square displacement relationship.

As mentioned above, the Hausdorff derivative orders are directly related to the fractal properties of the porous media. Without loss of generality, considering the anisotropic fractal media with three fractal dimensionalities β_1, β_2, and β_3, the three-dimensional Hausdorff fractal diffusion equation can be generalized as [22]

$$\frac{\partial u}{\partial t^{\alpha}} = D\left[\frac{\partial}{\partial x^{\beta_1}}\left(\frac{\partial u}{\partial x^{\beta_1}}\right) + \frac{\partial}{\partial y^{\beta_2}}\left(\frac{\partial u}{\partial y^{\beta_2}}\right) + \frac{\partial}{\partial z^{\beta_3}}\left(\frac{\partial u}{\partial z^{\beta_3}}\right)\right]. \tag{6.3}$$

6.1.2 The fundamental solution of the Hausdorff derivative diffusion equation

It has been demonstrated that the fundamental solution of the Hausdorff diffusion equation is formulated as [31]

$$u^{*}(x,t) = t^{-\alpha/2}\frac{1}{2\sqrt{\pi D}}\exp\left(-\frac{x^{2\beta}}{4Dt^{\alpha}}\right). \tag{6.4}$$

Considering the fundamental solution underlies the concept of distance, equation (6.4) should be correctly reformulated as [22]

$$u^{*}(x,y,t,\tau) = \left(t^{\alpha} - \tau^{\alpha}\right)^{-1/2}\frac{H(t-\tau)}{(4\pi D)^{1/2}}\exp\left(-\frac{(x^{\beta} - y^{\beta})^2}{4D(t^{\alpha} - \tau^{\alpha})}\right), \tag{6.5}$$

where y and τ denote the space and time reference point, respectively, and H is the Heaviside function. Equation (6.5) reduces to equation (6.4) with the reference point $\tau = 0$ and $y = 0$.

However, equation (6.4) is limited to the one-dimensional case. Considering varying constants β for all directions in equation (6.5), the fundamental solution for the three-dimensional Hausdorff derivative diffusion equation should be generalized as

$$u^*(x,y,t,\tau) = \left(t^\alpha - \tau^\alpha\right)^{-n/2} \frac{H(t-\tau)}{(4\pi D)^{n/2}} \exp\left(-\frac{r^{2\beta}}{4D(t^\alpha - \tau^\alpha)}\right), \tag{6.6}$$

where n represents the number of dimensions and r^β is defined in equation (2.11). It should also be mentioned that equation (6.6) is the fundamental solution of the anisotropic Hausdorff diffusion equation (6.3) with the distance r^β. It is obvious that the two-norm space-fractal distance and one-norm time-fractal distance are inherent in the fundamental solution.

It is noted that the fundamental solution equation (6.4) underlies the stretched Gaussian distribution from the viewpoint of space, and the Kohlrausch–Williams–Watts (KWW) stretched exponential relaxation function in terms of time.

6.1.3 Hausdorff advection–dispersion equation

Based on the Hausdorff derivatives, the time-space advection–dispersion equation (ADE) in one dimension is written as

$$\frac{\partial c(x,t)}{\partial t^\alpha} = \frac{\partial}{\partial x^\beta}\left(D(x,t)\frac{\partial c(x,t)}{\partial x^\beta}\right) - \frac{\partial}{\partial x^\beta}[Vc(x,t)], \tag{6.7}$$

where $c(x,t)$ is the solute concentration, i. e., the probability density function of the propagator located in an infinitesimal neighborhood dx centered in the distance x at time t; D is the diffusion (or dispersion) coefficient, and V is the flow velocity which could also be a function of the travel distance and time.

Here we present one form of the solutions with a constant diffusion coefficient and constant velocity. With $c = c(x,t)$ for simplicity and the pair of transformations

$$\begin{cases} \hat{t} = t^\alpha, \\ \hat{x} = x^\beta, \end{cases} \tag{6.8}$$

equation (6.7) becomes

$$\frac{\partial c}{\partial \hat{t}} = D\frac{\partial^2 c}{\partial \hat{x}^2} - V\frac{\partial c}{\partial \hat{x}}, \tag{6.9}$$

which can be transformed to the standard diffusion equation or heat equation with the aid of the Fuerth transform [96],

$$\frac{\partial u}{\partial \hat{t}} = D\frac{\partial^2 u}{\partial \hat{x}^2}. \tag{6.10}$$

With an instantaneous source input, i. e., a Dirac delta function input, $\delta(\hat{x})$, the solution of equation (6.10) is well known, i. e.,

$$u(\hat{x},\hat{t}) = \frac{1}{(4\pi D\hat{t})^{1/2}} \exp\left(-\frac{\hat{x}^2}{4D\hat{t}}\right); \tag{6.11}$$

$u = \delta(\hat{x})$ when $\hat{t} = 0$ and $u \to 0$ when $\hat{t} \to \infty$.

Then equation (6.11) can be retransformed to [114]

$$c(x,t) = \frac{M}{(4\pi Dt^{\alpha})^{1/2}} \exp\left[\frac{1}{2D}\left(Vx^{\beta} - \frac{x^{2\beta}}{2t^{\alpha}} - \frac{V^2 t^{\alpha}}{2}\right)\right]. \tag{6.12}$$

Equation (6.12) is the solution of the Hausdorff time-space fractal model with a constant diffusion coefficient and constant velocity. M is the total material released at the origin. The dimensions of D and V in equation (6.12) are $[L^{2\beta}/T^{\alpha}]$ and $[L^{\beta}/T^{\alpha}]$, respectively, where L is the unit of length and T is the unit of time. Bear in mind that equation (6.12) is the solute concentration in the Euclidean space with the Hausdorff fractal space-time parameters, β and α.

The connections between the Hausdorff fractal space and time are given by equation (6.8), and to express the dimensions of D and V in equation (6.12) in the Hausdorff fractal space and time, equation (6.8) is used to yield the dimensions of D and V as $[D] = [L^2/T]$ and $[V] = [L/T]$. Equation (6.12) is then equivalently written as

$$c(\hat{x},\hat{t}) = \frac{M}{(\pi D\hat{t})^{1/2}} \exp\left[\frac{1}{2D}\left(V\hat{x} - \frac{\hat{x}^2}{2\hat{t}} - \frac{V^2\hat{t}}{2}\right)\right]. \tag{6.13}$$

When $\beta = 1$ and $\alpha = 1$, the Hausdorff dimensions coincide with the Euclidean dimension so that the conventional dimensions of D and V in equation (6.13) become $[L^2/T]$ and $[L/T]$, respectively.

The advantage of equation (6.12) over equation (6.13) for parameter estimation is that data from the field or laboratories which are defined in the Euclidean space can be used to estimate the Hausdorff fractal space-time parameters, β and α. However, the final data on D and V need to be interpreted using equation (6.13) due to the fact that the dimensions of D and V in the Hausdorff space and time are $[L^2/T]$ and $[L/T]$, respectively. In other words, we use equation (6.12) to estimate the Hausdorff fractal parameters in the Euclidean space while interpreting their values in the Hausdorff space and time.

6.1.4 Numerical experiments

In this section, we focus on the time Hausdorff and the time fractional diffusion equations [45], which are known to describe sub-diffusive processes. The time Hausdorff derivative model is given by

$$\begin{cases} \dfrac{\partial u(x,t)}{\partial t^{\alpha}} = \dfrac{\partial^2 u(x,t)}{\partial x^2}, & 0 < x < L, t > 0, 0 < \alpha \le 1, \\ u(0,t) = u(L,t) = 0, \\ u(x,t) = \sin \dfrac{x\pi}{L}. \end{cases} \tag{6.14}$$

The time fractional diffusion equation is stated as

$$\begin{cases} \dfrac{d^{\gamma} u(x,t)}{dt^{\gamma}} = \dfrac{\partial^2 u(x,t)}{\partial x^2}, & 0 < x < L, t > 0, 0 < \gamma \le 1, \\ u(0,t) = u(L,t) = 0, \\ u(x,0) = \sin \dfrac{x\pi}{L}. \end{cases} \tag{6.15}$$

In equation (6.15), $d^{\gamma}u(x,t)/dt^{\gamma}$ represents the fractional derivative of the Caputo definition.

Here we used the Crank–Nicholson finite difference scheme for the above two models. The results are illustrated in Figure 6.2. From Figure 6.2, we find that the diffusion velocity in the time fractional derivative model is much larger than that of the time Hausdorff derivative model in the interval $t \in (0, 10]$. In the time Hausdorff model, the larger the Hausdorff derivative order α, the larger the diffusion velocity. In contrast, in the time fractional model, the larger the order γ, the slower the diffusion rate change in the initial interval $t \in (0, 1]$. But in $t \in (2, 10]$, the diffusion rate increases with the increasing γ. The time fractional derivative model exhibits a heavier tail than the time

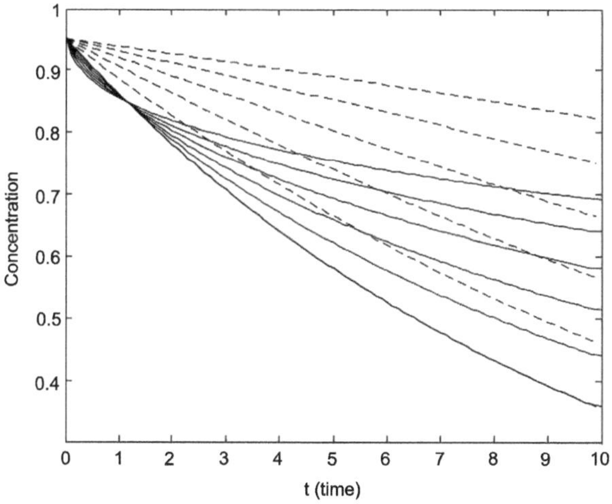

Figure 6.2: Diffusion image at $x = 0.6$; solid line: time Hausdorff model with different orders, from top to bottom: $\alpha = 0.5, 0.6, 0.7, 0.8, 0.9, 1.0$; dashed line: time fractional model with different orders, from top to bottom: $\gamma = 0.5, 0.6, 0.7, 0.8, 0.9, 1.0$.

Hausdorff derivative model. In addition, if we employ the space Hausdorff derivative model, it will exhibit the remarkable skewness property in the spatial domain.

6.2 Turbulence

Normal diffusion process in molecular scale and anomalous diffusion process in vortex scale are the two main factors considered in practical simulation. The power-stretched Gaussian distribution model is obtained based on these two scales [46] to fit the fully developed turbulence.

6.2.1 Normal diffusion process at the molecular scale

At the molecular scale of turbulence, the collision and friction between particles can be expressed by a stress variable which determines normal diffusion at the molecular scale. Considering normal diffusion circumstances, the probability density function of velocities at the molecular scale can be described by the following formula:

$$\frac{\partial u}{\partial t} = D_1 \nabla^2 u, \tag{6.16}$$

where D_1 is the diffusion coefficient. In the normal diffusion process, the velocities of random walkers obey Gaussian distribution.

6.2.2 Anomalous diffusion process at the vortex scale

At the vortex scale, turbulence diffusion exhibits heavy tail statistical distribution, an evident feature of anomalous diffusion. The coupling motion of large particles forms the vortex with fractal structures in different spatial and time scales. Therefore, anomalous diffusion at the vortex scale is the main factor leading to the random character of turbulence [5]. The fractional derivative and the fractal derivative are two important modeling tools for anomalous diffusion. However, the fractional derivative is inappropriate to analyze anomalous diffusion in turbulence by numerical simulation. As turbulence exhibits a fractal character, the Hausdorff derivative model has the ability to simulate turbulence anomalous diffusion. We have

$$\frac{\partial u}{\partial t} = D\left(\frac{\partial u}{\partial x^{\beta/2}}\left(\frac{\partial u}{\partial x^{\beta/2}}\right)\right), \tag{6.17}$$

where $0 < \beta \leq 2$ is the order of the Hausdorff derivative. The solution of equation (6.17) is a stretched Gaussian distribution function. And thus the expression of velocity increments can be written as

$$p(a) = C_1/\sqrt{2\pi\sigma^\beta}\exp(-a^\beta/2\sigma^\beta), \tag{6.18}$$

where σ is the variance and C the constant.

Here, we give the Hausdorff Laplace Reynolds equation [32]

$$\frac{\partial \bar{u}_i}{\partial t} + \bar{u}_j \cdot \frac{\partial \bar{u}_i}{\partial x_j} = -\frac{1}{\rho}\nabla \bar{p} + \upsilon \Delta \bar{u}_i + \kappa_0 \Delta_F^{1/3} \bar{u}_i, \tag{6.19}$$

where $\kappa_0 \Delta_F^{1/3} \bar{u}$ is the Laplace operator with fractal derivative.

6.2.3 Existing models

6.2.3.1 Lévy stable distribution model

Lévy stable distribution [116] has been used in many turbulence models. Compared with the Gaussian distribution, it can better describe a statistical process characterized by a heavy tail and it is also the solution of the fractional Reynolds equation [32]. The characteristic function of Lévy distribution is given as

$$\hat{l}_\alpha(\tau, k) = e^{-\tau |k|^\alpha}, \tag{6.20}$$

where $0 < \alpha \le 2$ denotes the Lévy stability index.

6.2.3.2 Tsallis distribution model

Tsallis distribution [13] has recently been used to describe the probability density function of Lagrangian turbulence particle accelerations. Its physical interpretation of turbulence is not satisfactory and related to the well-known Tsallis entropy. The Tsallis distribution has the following form when $q > 1$:

$$p_q(a) = \frac{1}{\delta}\left[\frac{q-1}{\pi(3-q)}\right]^{1/2} \frac{\Gamma(\frac{1}{q-1})}{\Gamma(\frac{3-q}{2(q-1)})} \frac{1}{[1 + \frac{q-1}{3-q}\frac{a^2}{\delta^2}]^{1/(q-1)}}, \tag{6.21}$$

where $|a| < \delta[(3-q)/(1-q)]^{1/2}$.

6.2.3.3 Stretched exponential distribution model

La Porta et al. [151] first provided a stretched exponential distribution; its parameters are given in the next section. Its form is

$$p(x) = C\exp(-x^2/((1 + |xa/c|)^b c^2)). \tag{6.22}$$

6.2.4 Power-stretched Gaussian distribution model

Molecular-scale normal diffusion and vortex-scale anomalous diffusion are considered the two types of essential mechanisms behind complex turbulence behaviors.

The cross-scale effect of these two scales determines the distribution shape of particle accelerations. The new probability density function, named power-stretched Gaussian distribution (PSGD), is given as [46]

$$p(a) = \frac{1}{\sqrt{2\pi\sigma^{\beta}}} \frac{1}{(|a| + a_0)^2} \exp\left(\frac{-(|a| + a_0)^{\beta}}{2\sigma^{\beta}} \right), \tag{6.23}$$

where a_0 is a constant.

6.2.5 Application on fully developed turbulence

In theoretical physics, to develop a multiscale model for fully developed turbulence is an open problem, which involves various spatial and temporal features covering many scales [165]. In 2001, La Porta et al. [151] did the most accurate measurement of particle transverse accelerations in fully developed turbulence using advanced technology. The experiment parameters were the following: $\tau_\eta = (\upsilon/\varepsilon)^{1/2}$ is the Kolmogorov time, where υ is the kinematic viscosity, ε is the turbulent energy dissipation, and η the Kolmogorov distance; $\tau_\eta = 0.93$ ms. This experiment successfully showed the three-dimensional, time-resolved trajectory of a tracer particle undergoing accelerations in violently turbulent water flow. In 2004, Mordant et al. [142] refined parameters as follows $a = 0.513 \pm 0.003$, $b = 1.600 \pm 0.003$, $c = 0.563 \pm 0.02$, $C = 0.733$, $R_\lambda = 690$. When $|x| \to \infty$, $p(x) \propto \exp(-|x|^{0.4})$.

But this statistical model has the obvious drawback of requiring the three obscure parameters, whose physical interpretation is missing. Besides, the model is simply a phenomenological description. To overcome this drawback, in 2001, Beck et al. [14] employed Tsallis distribution to refit the experimental data via the Tsallis entropy. Later on, Arimitsu et al. [5] proposed a multifractal model via the concept of nonextensive statistical mechanics, which can fit the experimental data accurately, but with complex expression. Above all, the physical mechanism behind the experimental data is still unclear in these models.

We used the fractional diffusion model presented by Chen et al. [46] based on the turbulence multiscale property and made a comparative study of these models. Table 6.1 provides the comparison of the Lévy stable distribution, the Tsallis distribution,

Table 6.1: Comparison of the four models.

Model	Parameters	Physical interpretation	Mean square error
Lévy stable distribution	2	accelerated diffusion	1.3972e3
Tsallis distribution	2	Tsallis entropy	19.9716
Stretched exponential	3	/	3.3373
PSGD	2	multiscale	3.5695

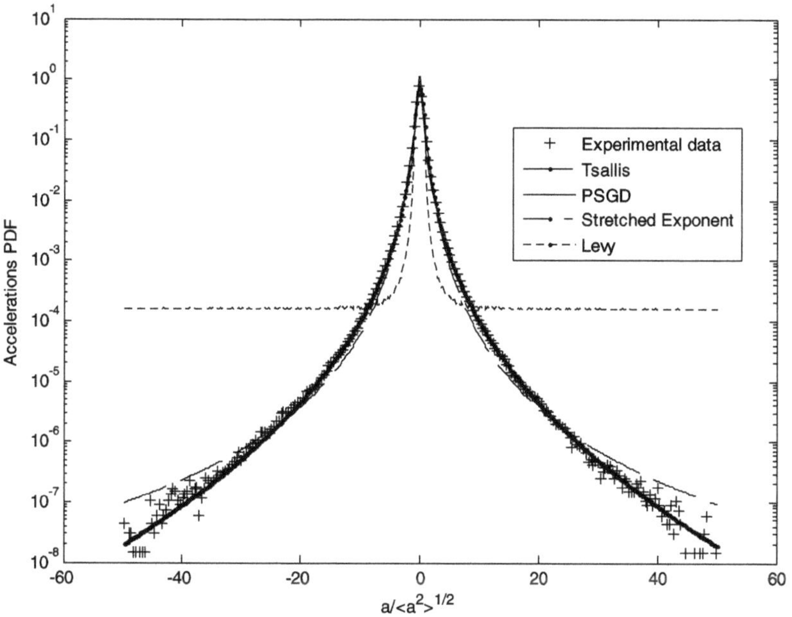

Figure 6.3: Lagrangian accelerations probability density function at $R_\lambda = 690$. Lévy fit with parameters: $R_\lambda = 1.9$, $a = 0.1$; Tsallis fit with parameters: $q = 1.5$, $\delta = 0.527$; power-stretched Gaussian fit with parameters: $\beta = 0.5401$, $\sigma = 0.2211$, coordinate offset: $x_0 = 0.3467$; stretched exponential fit with parameters: $a = 0.513$, $b = 1.600$, $c = 0.563$.

the stretched exponential distribution, and the PSGD in fitting the Lagrangian accelerations probability density function. Figures 6.3 and 6.4 give the fitting results of the Lagrangian accelerations probability density function and its fourth-order moment, respectively.

The Lévy stable distributions have been successfully used to describe chaos-induced turbulent diffusion, but Lévy stable distribution is not suitable to describe turbulence particle accelerations for its too slow decay rate, which deviates much from experimental data. The Tsallis distribution model was exploited by Beck [13] based on the distribution. It reveals the relationship between statistical theory of turbulence and non-extensive entropy, but this physical interpretation is not satisfactory. We find from Figure 6.3 that the Tsallis distribution does not fit experimental data either. It is stressed that the fourth-order moment of the Tsallis fitting differs dramatically from the experiment one.

It is noted that the stretched exponential distribution model, which is the first statistical model to fit the distribution of turbulence particle accelerations, makes an excellent fitting and results in the smallest error rate. The downside is that this model requires the three free parameters which lack an explicit physical interpretation. The model is simply a phenomenological description.

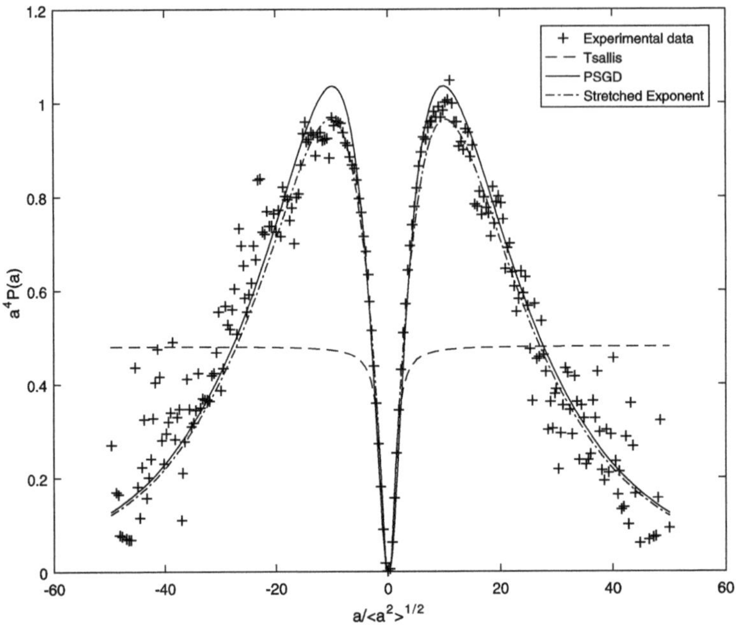

Figure 6.4: Comparison of the experimental accelerations probability density function at the fourth-order moment, where the parameters are the same as those in Figure 6.3.

The power-stretched Gaussian model is built on the cross-scale coupling of molecular-scale normal diffusion and vortex-scale anomalous diffusion. From Table 6.1 and Figures 6.3 and 6.4, we can clearly see that it fits the data very well. In addition, the model only requires two free parameters and has clear physical meaning. However, its connection to the Navier–Stokes equations is still tenuous. Therefore, the present approach and model are still under further investigation.

6.3 Viscoelasticity

A wide range of materials, such as rubber, soft soil, blood, colloid, and polymers, are observed to simultaneously exhibit both elastic and viscous behaviors [23]. It is well known that the stress in purely elastic and viscous processes is linear to the strain and the rate of strain, respectively. In contrast, the viscoelastic stress response of viscoelastic materials is dependent on time and strain rate, leading to complex behaviors of creep and relaxation in the case of certain stress or constraints. Such power law responses are usually considered as memorial behaviors, in other words, history-dependent process.

In recent decades, the viscoelastic mechanics models have undergone quick development. The differential and integral constitutive relationships have both been applied to the characterization of viscoelastic behaviors [52], i. e.,

$$\sum_{k=0}^{m} p_k \frac{d^k \sigma}{dt^k} = \sum_{k=0}^{n} q_k \frac{d^k \varepsilon}{dt^k}, \quad n \geq m, \tag{6.24}$$

$$\varepsilon(t) = \int_{-\infty}^{t} J(t-\zeta)\dot{\sigma}(\zeta)d\zeta, \tag{6.25}$$

where ε and σ denote strain and stress, respectively, p_k and q_k represent various materials constants, and $J(t)$ is the creep compliance. The integral constitutive model (6.25) corresponds to the Boltzmann superposition principle, while the differential model (6.24) is related to the spring-dashpot models.

It is often observed in experimental and field measurements that the classical viscoelastic models, e. g., the Maxwell, the Kelvin, the Burgers, and the standard linear solid models, fail to describe viscoelastic behaviors of complex materials, such as soft matter and fractal materials, especially for those characterizing the stretched exponential relaxation [139, 148].

In order to accommodate these so-called abnormal behaviors, within the classical viscoelastic models, more springs or dashpots have been employed to better fit the experimental data, which, however, has brought in some cumbersome parameters without clear physical significance. In addition, the applicability of these modified models is limited. Two typical models along this line are the generalized Maxwell and Kelvin models [197].

6.3.1 Fractal dashpot and fractal viscoelastic models

The fractal dashpot characterized by $\sigma_f = \eta d\varepsilon_f / dt^p$ is proposed by analogy with the Scott-Blair element of fractional derivatives, of which the schematic diagram is displayed in Figure 6.5 [23].

Figure 6.5: The schematic diagram of the fractal dashpot.

Figure 6.6 reveals the comparison of creep compliance between the fractal dashpot and the fractional Scott-Blair element [162]. It can be observed from Figure 6.6 that the creep compliance increases with increasing fractal orders. Moreover, the similar tendency of the curves indicates the capability of the fractal dashpot to describe the power law behavior.

By connecting a fractal dashpot in series and in parallel with a spring element, the fractal Maxwell and Kelvin model can be achieved, respectively. By analogy with

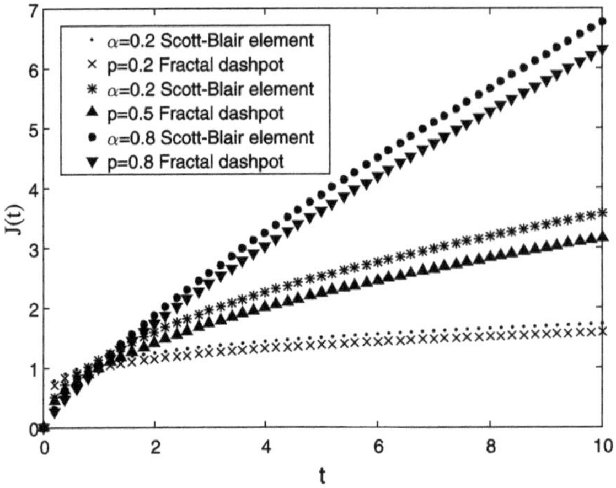

Figure 6.6: Comparison of creep compliance between the fractal dashpot and the Scott-Blair element.

the classical integer-order and the fractional Maxwell models, the basic equations of the fractal Maxwell model are formulated as

$$\begin{cases} \sigma_e = E\varepsilon_e, \quad \sigma_f = \eta \dfrac{d\varepsilon_f}{dt^p}, \\ \varepsilon = \varepsilon_e + \varepsilon_f, \quad \sigma = \sigma_e = \sigma_f, \end{cases} \tag{6.26}$$

where the subscripts e and f represent the elastic and the fractal elements, respectively, and E and η are Young's modulus and the viscosity coefficient, respectively. With the help of the stress–strain relation equation (6.26), it is straightforward to achieve the constitutive equation of the fractal Maxwell model as

$$\frac{d\varepsilon}{dt^p} = \frac{1}{E}\frac{d\sigma}{dt^p} + \frac{\sigma}{\eta}. \tag{6.27}$$

Similarly, it is not a difficult task to obtain the constitutive relationship equation of the fractal Kelvin and Zener model in the same manner, that is,

$$\frac{d\varepsilon}{dt^p} + \frac{E}{\eta}\varepsilon = \frac{\sigma}{\eta}, \tag{6.28}$$

$$E_1E_2\varepsilon + E_2\eta\frac{d\varepsilon}{dt^p} = (E_1 + E_2)\sigma + \eta\frac{d\sigma}{dt^p}, \tag{6.29}$$

where E_1 and E_2 are two different Young's moduli of the fractal Zener model shown in Figure 6.7.

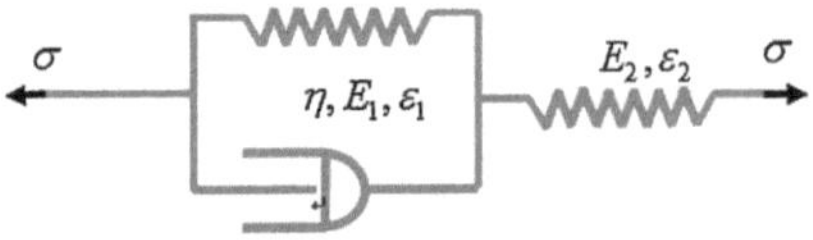

Figure 6.7: Schematic diagram of the fractal Zener model.

6.3.2 Scaling transformation

Here, two approaches are provided to derive the creep compliance and relaxation modulus for the proposed fractal models: rigorous derivation and scaling transformation [21]. Moreover, the achieved results are compared with those from classical integer-order and fractional models.

According to the intrinsic relationship of fractal derivatives

$$\frac{du}{dt^p} = \frac{1}{pt^{p-1}}\frac{du}{dt}, \tag{6.30}$$

equation (6.27) can be reformulated as

$$\frac{\dot{\varepsilon}}{pt^{p-1}} = \frac{1}{E}\frac{\dot{\sigma}}{pt^{p-1}} + \frac{\sigma}{\eta}. \tag{6.31}$$

In order to derive the creep compliance, the stress should be set as a constant, and the above equation can be rewritten as

$$\frac{\dot{\varepsilon}}{pt^{p-1}} = \frac{\sigma}{\eta}. \tag{6.32}$$

Applying the integration on both sides of equation (6.32) yields

$$\int \dot{\varepsilon}dt = \int \frac{\sigma}{\eta}pt^{p-1}dt. \tag{6.33}$$

Finally, by incorporating the initial value $\varepsilon_0 = \sigma/E$, the formulation of creep compliance for the fractal Maxwell model can be obtained in the following form:

$$J(t) = \frac{1}{E} + \frac{t^p}{\eta}. \tag{6.34}$$

The relaxation modulus can be achieved in the same manner as follows:

$$G(t) = Ee^{-t^p E/\eta}. \tag{6.35}$$

It is evident that the relaxation compliance of the fractal Maxwell model is in the similar form as the well-known stretched exponential relaxation.

Table 6.2 displays the derived creep compliance and relaxation modulus compared with those of the classical integer-order and fractional models. It is observed that the fractal and the fractional Maxwell models have almost the same creep modulus. It can also be seen in Table 6.2 that the creep modulus of the classical integer-order Kelvin model is a special case of the fractal Kelvin model.

It is noted that equations (6.34) and (6.35), respectively, can be directly obtained from the creep compliance and the relaxation modulus of the classical Maxwell model with the methodology of the scaling transformation $\hat{t} = t^p$ as shown in Table 6.2. Thus, the parameter p is supposed to be an inherent property parameter underlying the fractal of a given material. In this way, the creep compliance and relaxation modulus for the fractal models are easily achieved.

Table 6.2: Comparison between fractal, classical integer-order, and fractional derivative models [23].

		Integer-order derivative	Fractal derivative	Fractional derivative
Maxwell	Creep	$J(t)=\frac{1}{E}+\frac{t}{\eta}$	$J(t)=\frac{1}{E}+\frac{t^p}{\eta}$	$J(t)=\frac{1}{E}+\frac{t^\alpha}{\eta\Gamma(1+\alpha)}$
	Relaxation	$G(t)=Ee^{-tE/\eta}$	$G(t)=Ee^{-t^pE/\eta}$	$G(t)=EE_{\alpha,1}(-\frac{E}{\eta}t^\alpha)$
Kelvin	Creep	$J(t)=\frac{1}{E}(1-e^{-Et/\eta})$	$J(t)=\frac{1}{E}(1-e^{-Et^p/\eta})$	$J(t)=\frac{1}{\eta}t^\alpha E_{\alpha,\alpha+1}(-\frac{E}{\eta}t^\alpha)$
	Relaxation	$G(t)=E+\eta\delta(t)$	$G(t)=E+\eta\delta(t^p)$	$G(t)=E+\eta\frac{t^{-\alpha}}{\Gamma(1-\alpha)}$

6.3.3 Characterizing anomalous rheological behaviors with scaling transformation

Viscoelastic materials always exhibit features of both elastic solid and viscous fluid. The phenomenological behaviors of such materials can be illustrated by models consisting of elastic springs and viscous dashpots in parallel or in series. The three most famous fundamental models are known as the Maxwell, Kelvin, and Zener models. The Maxwell and Kelvin models are, respectively, employed to describe relaxation and creep behaviors on the basis of the various material properties, while the Zener model is suitable for both cases.

It has been found that these fundamental models are not capable of characterizing complex rheological behaviors [159] which exhibit a heavy-tailed feature. In order to overcome such problems, more elements, either spring or dashpot, have been added to those fundamental models, resulting in a more complex model with cumbersome parameters.

In order to facilitate such troublesome modeling, this section focuses on forging links between classical rheological models and anomalous viscoelastic behaviors with t^α law as revealed by the relationship between normal and anomalous diffusion.

As is well known, the Maxwell model is suitable for describing relaxation behavior, and the corresponding relationship between relaxation time and stress is

$$\sigma_A = a\exp(-t/b), \tag{6.36}$$

where a and b are constants and σ represents stress. Equation (6.36) is in agreement with the Debye exponential relaxation function. Equation (6.36) can be modified with the scaling transformation in the form of

$$\sigma_B = a\exp(-t^\alpha/b), \tag{6.37}$$

where a, b, and α are constants.

Porcine cornea is a kind of biological tissue with distinct viscoelastic properties. Relaxation experiments with porcine cornea have been carried out at different ablation depths [60]. It can be observed from Figure 6.8 that the relaxation cannot be well described by the traditional Maxwell model. Equation (6.37) is found to better fit the experimental data than equation (6.36), as shown in Figure 6.8. To achieve the error

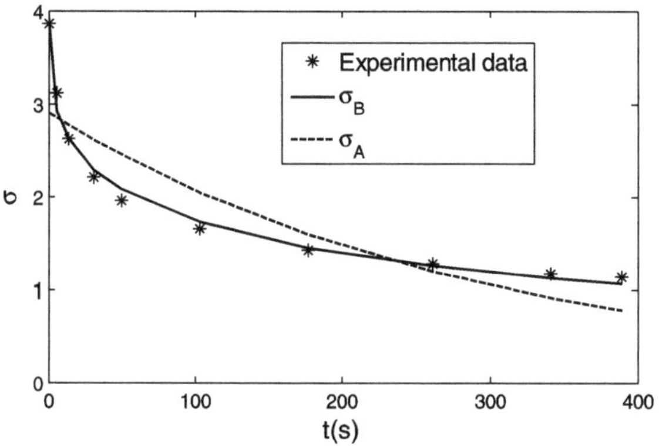

(a) 30% ablation depths

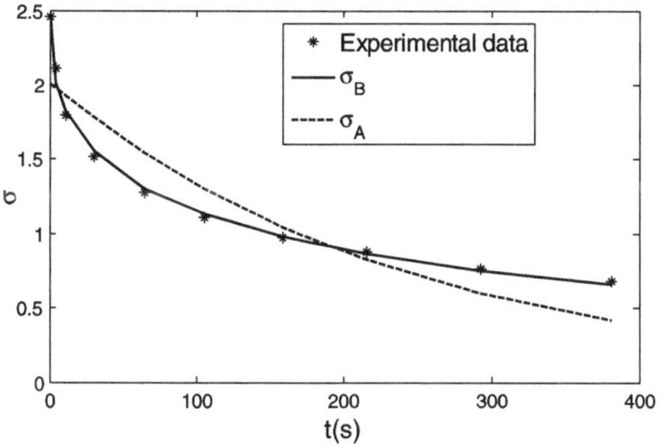

(b) 70% ablation depths

Figure 6.8: Comparison between experimental and numerical results for relaxation of porcine cornea (data adopted from reference [60]).

between experimental and fitting results, the average relative error is defined as

$$\text{Rerr} = \left(\sqrt{\frac{\sum_{k=1}^{n} (\sigma_{0k} - \sigma_k)^2}{\sum_{k=1}^{n} \sigma_k^2}} \right), \tag{6.38}$$

where σ_{0k} and σ_k represent the numerical and experimental data at different points k, respectively, and n denotes the number of experimental points. The average relative errors for the cases in Figure 6.8 are displayed in Tables 6.3 and 6.4, which in turn also validate the efficiency of the modified formulation. The results are further illustrated in Figure 6.9 with t^α as abscissa. It can be observed obviously from Figure 6.9 that equation (6.37) fits the experimental data better with only one more parameter α. The values of α vary at different ablation depths, as shown in Figures 6.8 and 6.9, which indicates that the viscoelastic properties of porcine cornea alter at different depths. The modified relaxation relationship is identical to the KWW stretched exponential law, i. e., equation (6.37).

Table 6.3: Best-fitting parameters and average relative error for the experimental test of porcine cornea at 30 % ablation depth.

	a	*b*	*α*	**Rerr**
σ_B	3.91	6.2	0.35	0.0378
σ_A	2.9	295		0.1914

Table 6.4: Best-fitting parameters and average relative error for the experimental test of porcine cornea at 70 % ablation depth.

	a	*b*	*α*	**Rerr**
σ_B	2.48	8.6	0.41	0.0250
σ_A	2.01	242.6		0.1534

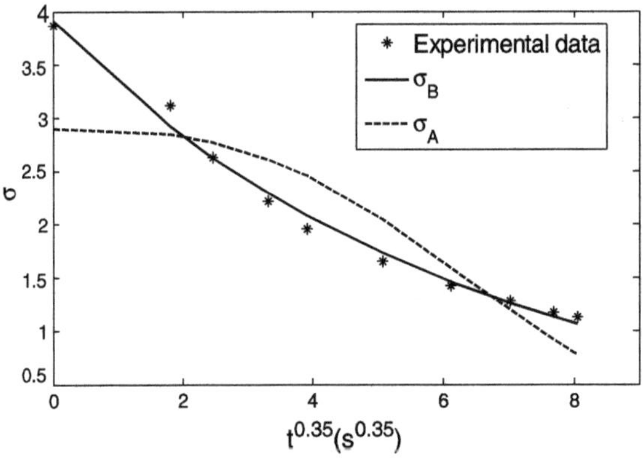

(a) 30% ablation depths

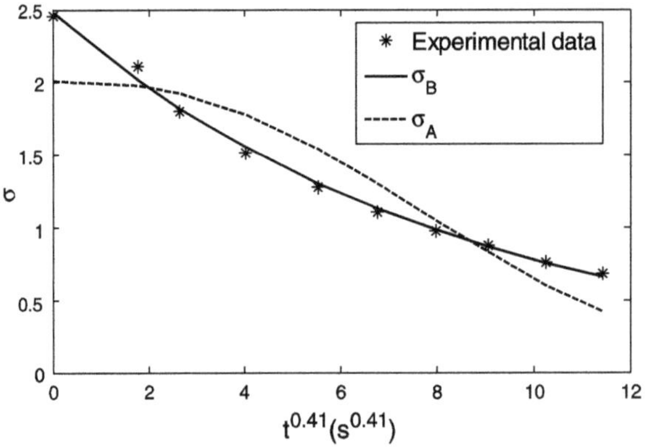

(b) 70% ablation depths

Figure 6.9: Comparison between experimental and numerical results for relaxation of porcine cornea with t^{α} as abscissa (data adopted from reference [60]).

If t goes to infinity in equation (6.37), the relaxation stress vanishes to 0, while for a large variety of viscoelastic materials, the stress will be a constant for a long time, known as plateau modulus, which plays an important role in the study of rheological behavior. Under such consideration, equation (6.37) should be written in the following form:

$$\sigma_D = a + b\exp(-t^{\alpha}/c), \tag{6.39}$$

where a, b, c, and α are constants. Without scaling transformation, equation (6.39) reduces to the following form:

$$\sigma_C = a + b\exp(-t/c). \tag{6.40}$$

Equation (6.40) is found to agree with the relaxation modulus of the classical Zener model. Subsequently, both equations are used to fit the experimental data. Without loss of generality, experimental data are quoted from three different viscoelastic materials, and they fall in different time scales, varying from 10^3 min to 10^6 min. For the sake of convenience, the fitting results are only illustrated in the figures, with t^{α} as abscissa, and Table 6.5 only displays the average relative error for different cases without the fitting parameters. It can be seen from Figure 6.10 and Table 6.5 that the classical relaxation relationship can well fit complex relaxation processes with scaling transformation.

Table 6.5: Average relative error for data fitting of various geomaterials.

	Rerr		
	Slip zone soil	**Saturated soil**	**Soft clay**
σ_D	0.0057	0.0148	0.0073
σ_C	0.0220	0.0294	0.0151

6.3.4 Complex creep behaviors with t^{α} law

Like the Maxwell model, the Kelvin model is utilized to describe creep behaviors, and its creep response can be modeled as

$$\varepsilon_A = a(1 - \exp(-t/b)), \tag{6.41}$$

where a and b are constants. Similarly, equation (6.41) also encounters the difficulty in characterizing creep behaviors of complex viscoelastic media, such as soil and rock. In this study, the t^{α} law modification is applied to the classical creep relationship. In doing so, the modified creep response is obtained as

$$\varepsilon_B = a(1 - \exp(-t^{\alpha}/b)), \tag{6.42}$$

where α is an additional parameter.

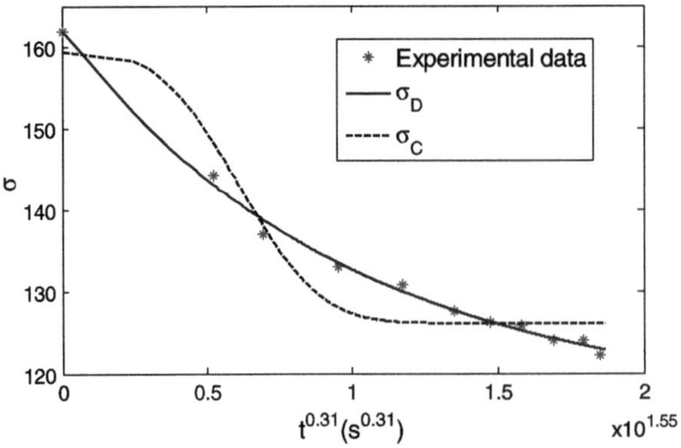

(a) Slip zone soil (data adopted from Ref. [190])

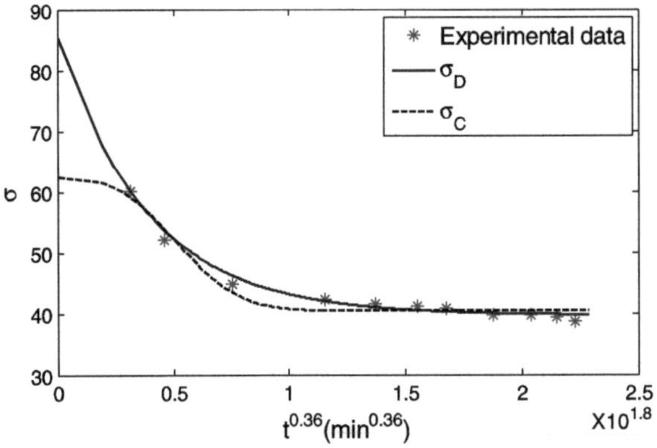

(b) Saturated Soil (data adopted from Ref. [202])

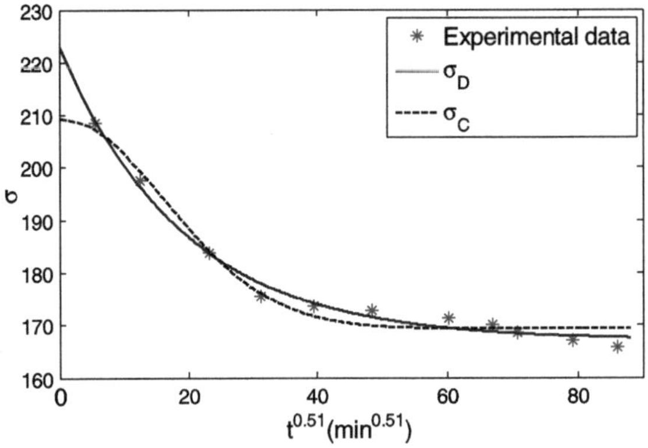

(c) Soft Clay (data adopted from Ref. [109])

Figure 6.10: Comparison between experimental and numerical results for relaxation of various geomaterials with t^{α} as abscissa.

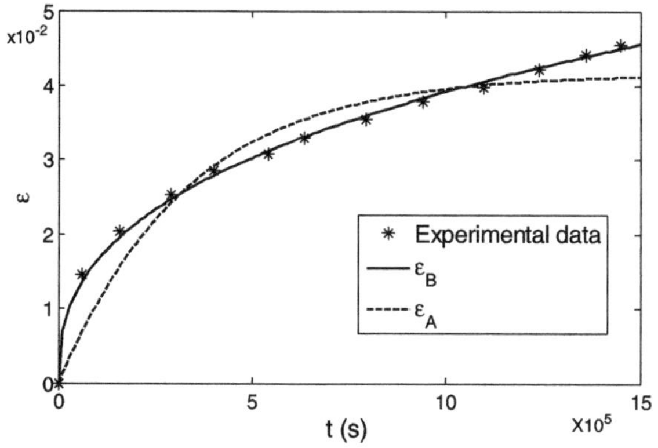

(a) The abscissa is t

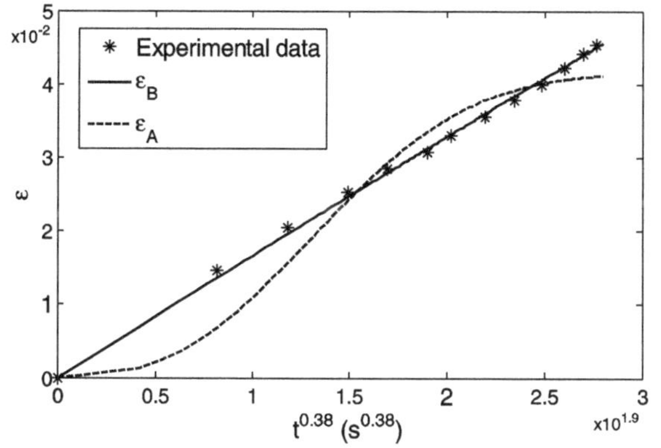

(b) The abscissa is t^{α}

Figure 6.11: Comparison between experimental and numerical results for creep of rock salt (data adopted from reference [54]).

Rock salt is usually regarded as a kind of viscoelastic material. Thus, a collection of experimental data of rock salt is used to examine the efficiency of the proposed formulation [54]. Figure 6.11(a) shows that the modified creep relationship fits the experimental data better than the classical one. The comparison is also illustrated in Figure 6.11(b) where the abscissa is t^{α}. Table 6.6 presents the fitting parameters and average relative error. The same observations can be made from Figure 6.11(b) and

Table 6.6: Best-fitting parameters and relative error for the experimental test of rock salt.

	a	b	α	Rerr
ε_B	0.07	0.6	0.38	0.0171
ε_A	0.24	0.8		0.0986

Table 6.6, i. e., the classical creep response is capable to describe the creep behavior better by using t^α modification.

Numerous experiments reveal that the creep response starts from an initial value other than 0. Thus, equation (6.41) should be modified to obey the following relationship:

$$\varepsilon_C = a + b(1 - \exp(-ct)), \tag{6.43}$$

where a, b, and c are the three constant parameters. Equation (6.43) corresponds to the creep compliance of the classical Zener model. Recalling the t^α law modification, the modified creep response can be reformulated as

$$\varepsilon_D = a + b(1 - \exp(-ct^\alpha)). \tag{6.44}$$

The experimental data for concrete B300 are examined to test the achieved creep response [90]. As shown in Figure 6.12, the abscissa is set to be $t^{0.5}$, and equation (6.44) displays a better data fitting. The average relative error from Table 6.7 confirms the advantage of the modified formulation.

According to the discussions above, it seems that the three fundamental viscoelastic models have revived with t^α law modification for power law viscosity, as the modified models fit experimental data well without cumbersome parameters. In this subsection, the purpose is to connect the achieved models to the existing fractal derivative

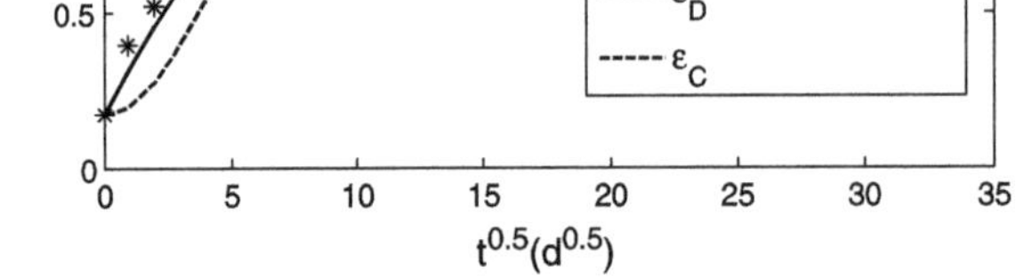

Figure 6.12: Comparison between experimental and numerical results for creep of concrete B300 (data adopted from reference [90]).

Table 6.7: Best-fitting parameters and relative error for the experimental test of concrete B300.

	a	b	c	α	Rerr
ε_D	0.17	1.475	0.109	0.5	0.0296
ε_C	0.17	1.289	0.0216		0.1287

modeling formalism. With the combination of such dashpot and spring in parallel or in series, the fractal derivative viscoelastic models have been established. According to reference [23], it is worth mentioning that equation (6.37) agrees with the relaxation modulus of the fractal derivative Maxwell model, while equation (6.42) corresponds to the creep compliance of the fractal derivative Kelvin models. Moreover, equations (6.39) and (6.44) can be regarded as the relaxation modulus and creep compliance of the fractal derivative Zener model, which consists of two elastic springs and one fractal dashpot. The fractal derivative Zener model is more suitable to describe complex rheological behaviors.

Traditionally, the viscoelastic models with a few more elements are often employed to characterize the creep behavior of complex materials, especially when the classical Zener model fails to meet the requirement of data fitting. Under such consideration, the viscoelastic models with five or more elements have been proposed [66]. The comparison of the classical Zener, the modified Zener, and the five-element models has been discussed in reference [23]. Table 6.8 presents the corresponding errors. It is observed that the modified Zener formulation achieves a higher fitting accuracy than either the classical Zener model or the five-element model. That is to say, the model with more than five elements is required to get the same accuracy as the modified Zener model. Thus, we can see that the modified formulation is more efficient in modeling such complex creep behaviors with fewer parameters.

Table 6.8: The average relative error of different models.

	Classical Zener model	Modified Zener model	Five-element model
Rerr	0.1287	0.0296	0.0609

The relaxation modulus and creep compliance of the fractal derivative models can be obtained directly with t^α modification, i. e., the methodology of scaling transformation. It is also noted that they can also be derived rigorously with the following relationship between the fractal derivative and ordinary first-order derivative [23]. It can be concluded that the derived formulations are directly linked to those achieved with t^α modification and the fractal derivative modeling formalism is inherent with scaling transformation.

6.3.5 Comparisons of different relaxation functions

In this subsection, we investigate and compare the dynamics of relaxation processes characterized by the fractal and the fractional derivative models. The classical relaxation equation, also known as Debye relaxation, is formulated as

$$\begin{cases} \dfrac{d\phi(t)}{dt} + \lambda\phi(t) = 0, \\ \phi(t=0) = \phi_0, \end{cases} \tag{6.45}$$

where λ is a constant coefficient. The analytical solution of equation (6.45) is given by

$$\phi(t) = \phi_0 e^{-\lambda t}. \tag{6.46}$$

Recently, the relaxation equation of the fractional derivative has also been proposed, which is given in [129], i. e.,

$$\frac{d^\beta\phi(t)}{dt^\beta} + \lambda\phi(t) = 0, \tag{6.47}$$

where β is the fractional derivative order. With the same initial condition as shown in equation (6.45), its analytical solution is derived in the following form:

$$\phi(t) = \phi_0 E_\beta(-\lambda t^\beta), \tag{6.48}$$

where $E_\alpha(\cdot)$ represents the Mittag-Leffler function.

By analogy with the classical Debye relaxation, the fractal derivative relaxation equation is straightforwardly written as

$$\begin{cases} \dfrac{d\phi(t)}{dt^\alpha} + \lambda\phi(t) = 0, \\ \phi(t=0) = \phi_0, \end{cases} \tag{6.49}$$

of which the analytical solution is given by

$$\phi(t) = \phi_0 e^{-\lambda t^\alpha}. \tag{6.50}$$

It is worth mentioning that the fractal derivative relaxation model is directly related to the well-known stretched exponential function. The comparisons between different relaxation models are illustrated in Figure 6.13. It is observed that both the fractal and the fractional derivative relaxation models agree well with the Debye relaxation when $\alpha = \beta = 1$ and are capable to describe the heavy-tailed phenomenon, though the fractal derivative relaxation model decays faster than the fractional one. With varying fractal derivative order, the curves intersect at $t = 1$.

6.3.6 Remarks

Studies find that the normal diffusion can be linked to the anomalous one with scaling transformation. Enlightened by such a relationship, the relaxation modulus of the classical Maxwell model is modified with t^α law. It is observed that the proposed formulation well fits the experimental data with high accuracy, and appears similar to

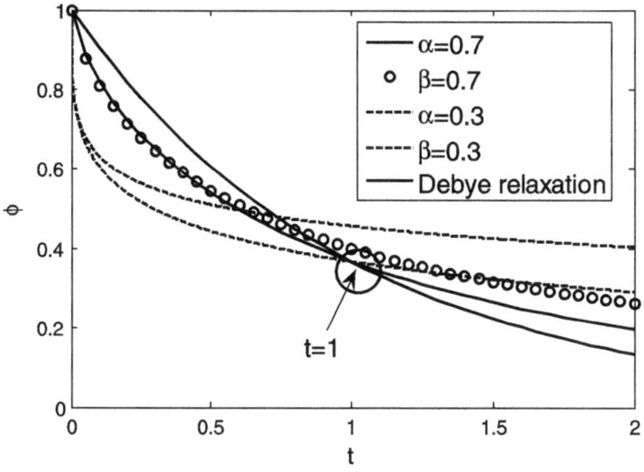

Figure 6.13: Comparison between different relaxation models.

the well-known KWW stretched exponential function. Such function has also been generalized to characterize complex relaxation behaviors whose limits deviate from 0. The efficiency of the present reformulation has been validated by experimental data at different time scales.

The same modification is also applied to the creep compliance. In comparison with the traditional models, the proposed models are observed to achieve higher fitting accuracy. In addition, the modified formulation only requires one more parameter α without cumbersome parameters. Thus, the classical Maxwell, Kelvin, and Zener viscoelastic models have been reactivated to describe complex rheological behaviors with the proposed methodology of scaling transformation.

The generalized relaxation and creep functions can also rigorously be derived from the fractal derivative Zener model, which is found capable to characterize either the creep behavior with an initial value, or the relaxation process whose limit deviates from 0. The rigorously derived creep compliance and relaxation modulus are also found in the same form as those modified with t^{α}. It has been further observed that the modified relaxation modulus of the Maxwell model and the creep compliance of the Kelvin model can be directly obtained from the corresponding fractal derivative Maxwell and Kelvin models. Thus, it can be concluded that the fractal derivative operator is inherently consistent with scaling transformation in time and is capable to describe anomalous rheological behaviors of complex viscoelastic media.

6.4 Creep and relaxation

6.4.1 Fractal models of creep and relaxation

A loading and unloading numerical experiment is conducted, as shown in Figure 6.14, where the stress remains stable from time $t_1 = 1$ to time $t_2 = 2$ and is then re-

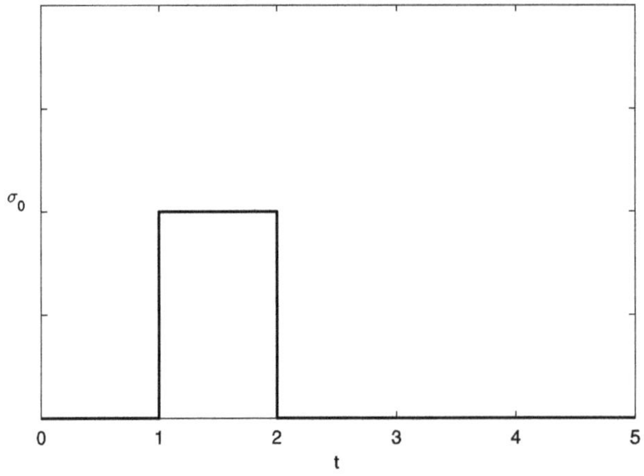

Figure 6.14: The stress in the loading and unloading processes.

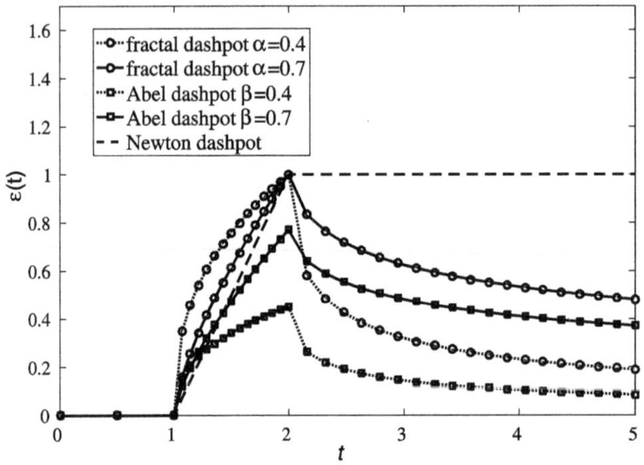

Figure 6.15: The creep and recovery curves of the fractal dashpot, Newton dashpot, and Abel dashpot.

moved. The results of creep and recovery for these three dashpots are shown in Figure 6.15 [174].

From Figure 6.15 we can see that, at a small value of α, the strain of the fractal dashpot changes quickly both on the loading and unloading processes, which exhibits the high elasticity of the fractal dashpot. On the other hand, with the order α approaching 1, the strain of the fractal dashpot is close to that of the Newton dashpot (a special case of the fractal dashpot with order equal to 1), which shows the high viscosity of the fractal dashpot. In addition, it can be seen that the fractal dashpot and Abel dashpot are almost similar in viscoelastic property with respect to the creep and recovery processes. The non-steady flow of non-Newtonian fluid obeying the fractal dashpot model was numerically studied in [175].

On the other hand, creeps of the fractal Maxwell and the fractional Maxwell models have a similar power law form. The relaxations of Maxwell on the Hausdorff derivative and the fractional derivative have a very close relation considering the following expression [136]:

$$E_\alpha(-\lambda t^\alpha) \sim \exp\left(-\frac{\lambda t^\alpha}{\Gamma(1+\alpha)}\right), \quad \lambda \to 0 \text{ or } t \to 0, \tag{6.51}$$

where $E_\alpha(\bullet)$ denotes the Mittag-Leffler function with one parameter.

Figures 6.16 and 6.17, respectively, give the creep curves of Maxwell and Kelvin models on three derivatives by letting $E = \eta = \mu = \xi = 1$ for convenience. Fig-

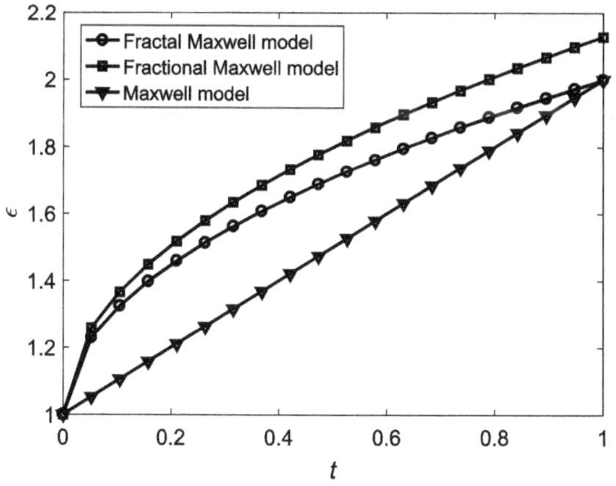

Figure 6.16: Creep of Maxwell, fractal Maxwell, and fractional Maxwell models ($\alpha = \beta = 0.5$).

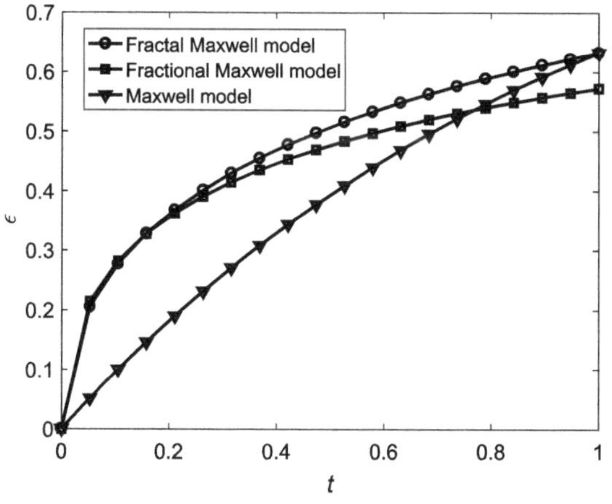

Figure 6.17: Creep of Kelvin, fractal Kelvin, and fractional Kelvin models ($\alpha = \beta = 0.5$).

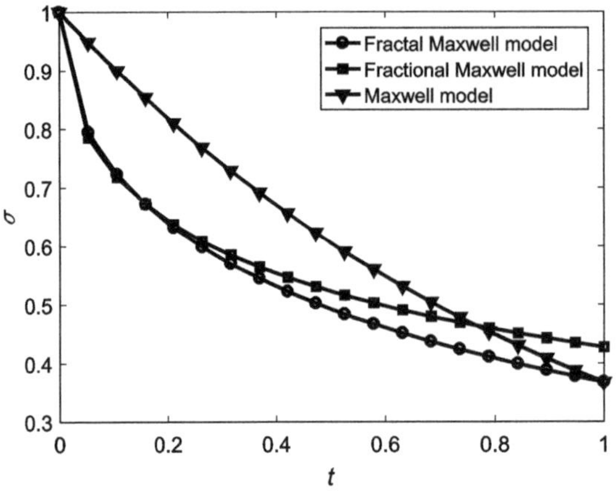

Figure 6.18: Relaxations of Maxwell, fractal Maxwell, and fractional Maxwell models ($\alpha = \beta = 0.5$).

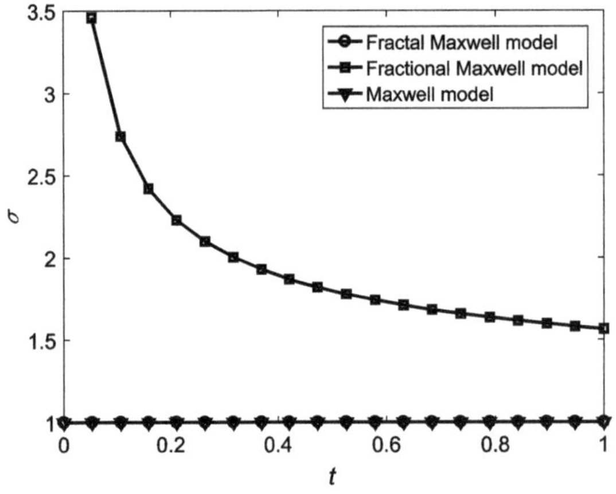

Figure 6.19: Relaxations of Kelvin, fractal Kelvin, and fractional Kelvin models ($\alpha = \beta = 0.5$).

ures 6.18 and 6.19 are the corresponding relaxation curves. Figure 6.16 shows that the fractal Maxwell model remedies terrible behavior of the Maxwell model in describing creep. Figures 6.17 and 6.18 indicate the fractal models have creep and relaxation responses similar to the corresponding fractional models. Figure 6.19 shows that the fractal Kelvin model cannot predict relaxation behavior of viscoelastic material.

Taking several fractal Kelvin models combination in series, the generalized fractal Kelvin model [21] is obtained, as shown in Figure 6.20. The constitutive relation obeys

$$\begin{aligned} \sigma &= E_0 \varepsilon_0, \\ \frac{d\varepsilon_i}{dt^{\alpha_i}} + \frac{E_i}{\eta_i}\varepsilon_i &= \frac{\sigma}{\eta_i}, \\ \varepsilon &= \varepsilon_0 + \sum_{i=1}^{n} \varepsilon_i. \end{aligned} \tag{6.52}$$

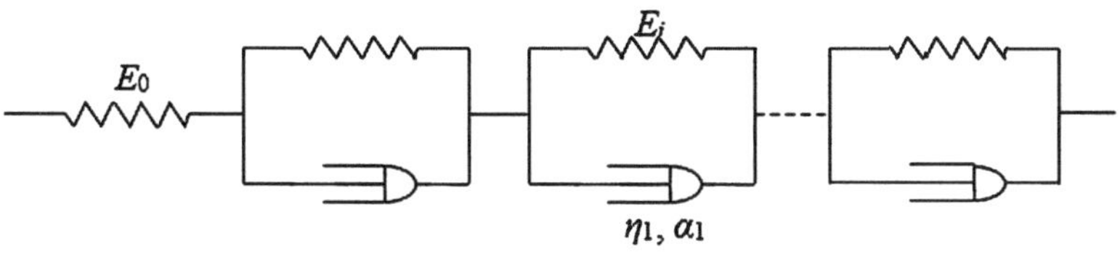

Figure 6.20: Generalized fractal Kelvin model.

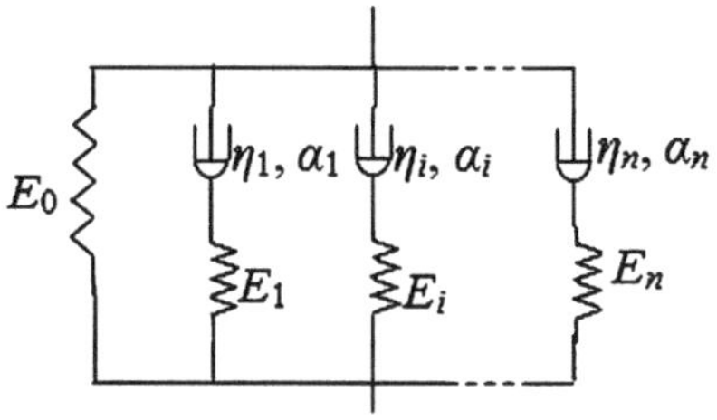

Figure 6.21: Generalized fractal Maxwell model.

Let $\sigma = \sigma_0$, the creep compliance can be written as

$$J(t) = \frac{1}{E_0} + \sum_{i=1}^{n} \left\{ \frac{1}{E_i} (1 - e^{-E_i t^{\alpha_i}/\eta_i}) \right\}. \tag{6.53}$$

Taking a combination of several fractal Maxwell models in parallel, the generalized fractal Maxwell model is obtained, as shown in Figure 6.21. The constitutive relation obeys

$$\begin{aligned} \sigma_0 &= E_0 \varepsilon, \\ \frac{1}{E_i} \cdot \frac{d\sigma_i}{dt^{\alpha_i}} + \frac{\sigma_i}{\eta_i} &= \frac{d\varepsilon}{dt^{\alpha_i}}, \\ \sigma &= \sigma_0 + \sum_{i=1}^{n} \sigma_i. \end{aligned} \tag{6.54}$$

Let $\varepsilon = \varepsilon_0$, the relaxation modulus can be written as

$$G(t) = E_0 + \sum_{i=1}^{n} E_i e^{-\frac{E_i}{\eta_i} \cdot t^{\alpha_i}}. \tag{6.55}$$

A dynamic load is set to further study the strain response characteristics of the fractal dashpot. For simplicity, the dynamic load is set as a form of sinusoidal function, i. e.,

$$\sigma(t) = \hat{\sigma} \sin 2\pi f t, \tag{6.56}$$

where $\hat{\sigma}$ is the amplitude and f is the frequency. The strain response of the fractal dashpot can be written as

$$\varepsilon(t) = \frac{\hat{\sigma} \cdot \alpha}{\eta} \cdot \int_0^t \tau^{\alpha-1} \sin 2\pi f \tau d\tau. \tag{6.57}$$

Similarly, the strain responses of the Newton dashpot and Abel dashpot are presented as

$$\varepsilon(t) = -\frac{\hat{\sigma}}{2\pi f \cdot \mu} \cdot \cos 2\pi f t \tag{6.58}$$

and

$$\varepsilon(t) = \frac{\hat{\sigma}}{\xi \cdot \Gamma(\beta)} \cdot \int_0^t \frac{\sin 2\pi f \tau d\tau}{(t-\tau)^{1-\beta}}. \tag{6.59}$$

For the sake of convenience, all the material parameters and the amplitude and frequency of stress are set as 1, i. e., $\eta = \mu = \xi = f = \hat{\sigma} = 1$. The strain responses of three dashpots are shown in Figures 6.22–6.24. From these three figures, we can find that the fractal dashpot has the same cycle and phase as the Abel dashpot and Newton dashpot. Although the amplitude of strain for the Newton dashpot remains stable, that for both the fractal dashpot and the Abel dashpot fades with time. Furthermore, the amplitude fades more quickly with smaller derivative order for both the fractal dashpot and the Abel dashpot. It is worth mentioning that the computational time of the fractal dashpot is much less than that of the Abel dashpot, as shown in Table 6.9, which means the fractal dashpot displays a higher computational efficiency compared with the Abel dashpot.

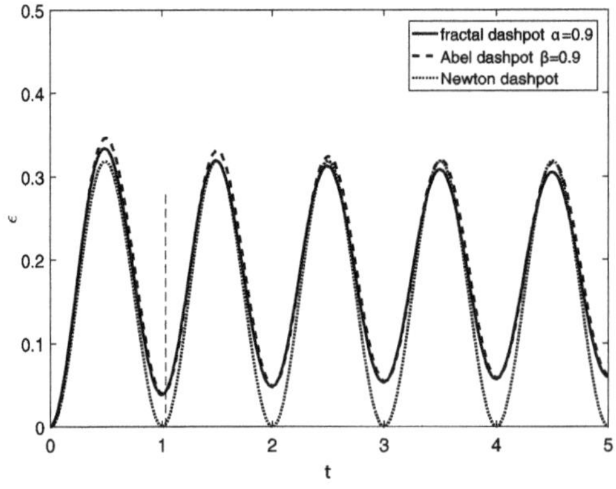

Figure 6.22: The strain responses of the fractal dashpot, Abel dashpot, and Newton dashpot under dynamic load at derivative order 0.9.

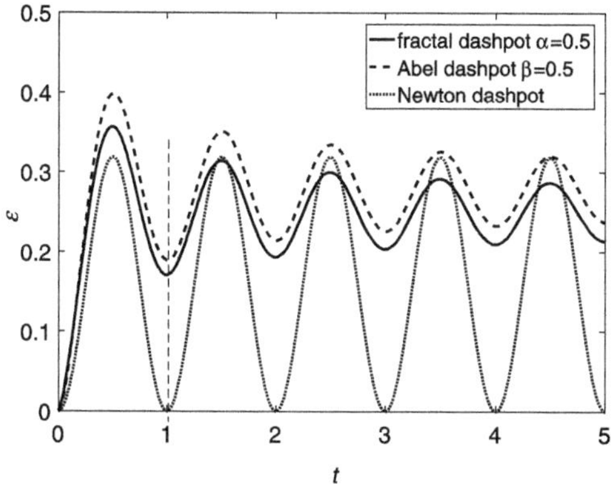

Figure 6.23: The strain responses of the fractal dashpot, Abel dashpot, and Newton dashpot under dynamic load at derivative order 0.5.

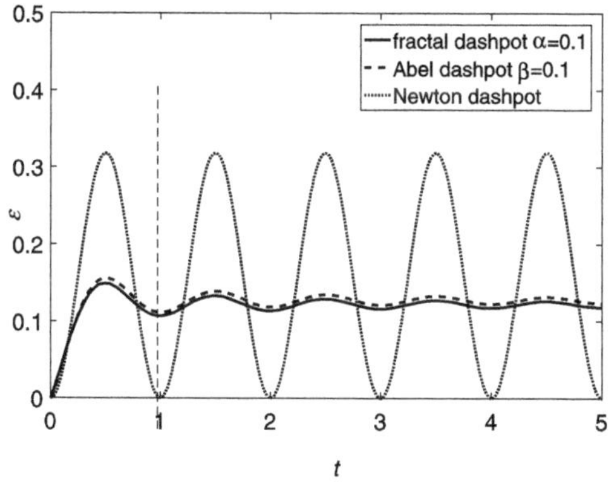

Figure 6.24: The strain responses of the fractal dashpot, Abel dashpot, and Newton dashpot under dynamic load at derivative order 0.1.

Table 6.9: The CPU time the for fractal dashpot and Abel dashpot (AMD A6-3420M, RAM 2.74GB, 32 bit Windows 7 and MATLAB 2010b).

	Fractal dashpot			Abel dashpot		
Derivative order	**0.1**	**0.5**	**0.9**	**0.1**	**0.5**	**0.9**
$\Delta t = 0.01s$	0.1258	0.1356	0.015067	0.5029	0.5004	0.501391
$\Delta t = 0.001s$	0.1333	0.1333	0.2240s	483.6862	483.4233	469.9336

Let the strain rate be a constant, i. e., $\dot{\varepsilon} = c$. The stress response of the fractal dashpot can be written as

$$\sigma = \frac{\eta \cdot c \cdot t^{1-\alpha}}{\alpha}. \tag{6.60}$$

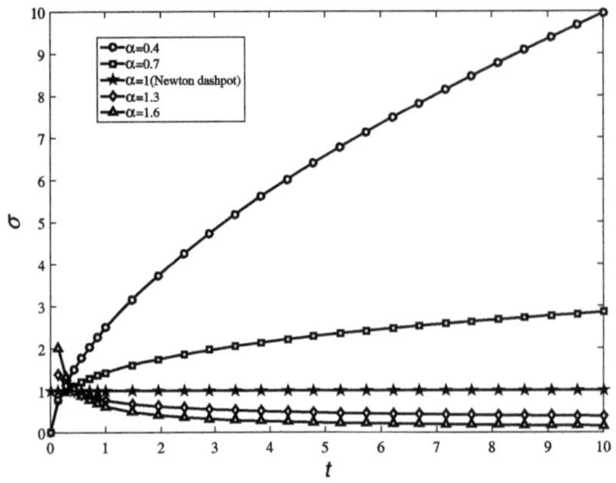

Figure 6.25: The stress response of the fractal dashpot at constant strain rate.

For the sake of convenience, the material parameter and strain rate are set as 1, i. e., $c = \eta = 1$. The stress response is shown in Figure 6.25.

The apparent viscosity [85] of non-Newtonian fluid is generally defined by the ratio of stress to strain rate. In this situation, the apparent viscosity can be written as

$$\eta_b = \frac{\eta \cdot t^{1-\alpha}}{\alpha}. \tag{6.61}$$

It is easy to find that the stress response under constant strain rate is a form of power law function. From Figure 6.25, the stress is found to decrease with time when the derivative order is bigger than 1 and it keeps increasing when the order is less than 1. Apparent viscosity changes the same way as the stress response, which means in this situation the fractal dashpot is suitable to characterize the time-dependent viscosity for non-Newtonian fluid, including "rheopexy" [147] and "thixotropy" [10] phenomena.

The strain rate is set as a linear function of time, i. e., $\dot{\varepsilon} = c \cdot t$; the stress response of the fractal dashpot can be written as

$$\sigma = \frac{\eta \cdot c^{\alpha-1}}{\alpha} \cdot \dot{\varepsilon}^{2-\alpha}. \tag{6.62}$$

Figure 6.26 shows the stress response with strain rate ($\eta = c = 1$).

From Figure 6.26, we can see that the stress is shown as a lower convex function when $0 < \alpha < 1$ and an upward convex function when $\alpha > 1$. That agrees with the shear thickening and shear thinning phenomena [81] for non-Newtonian fluid.

In general, the stress response of the fractal dashpot under power law strain rate ($\dot{\varepsilon} = c \cdot t^n$) can be written as

$$\sigma = \frac{\eta}{\alpha} \cdot c^{\frac{1}{n}\cdot(\alpha-1)} \cdot \dot{\varepsilon}^{1+\frac{1}{n}\cdot(1-\alpha)}. \tag{6.63}$$

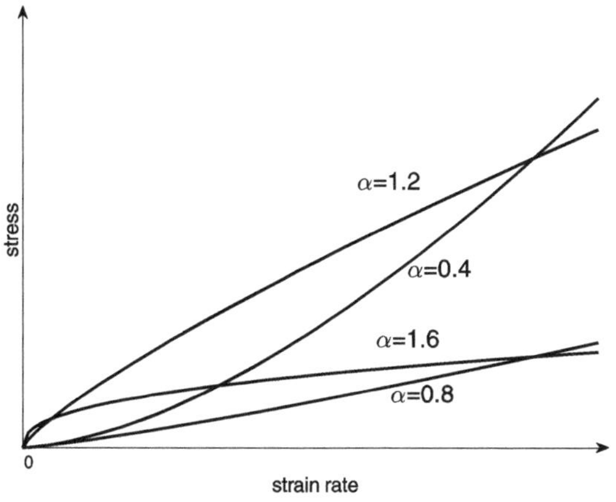

Figure 6.26: The stress response of the fractal dashpot at time-linear strain rate.

Equation (6.63) can be used to characterize the shear thickening phenomenon with $\alpha < 1$ and the shear thinning phenomenon with $\alpha > 1$.

A dynamic strain is set to further study the stress response characteristics of the fractal dashpot. Similarly, we set the strain as a sinusoidal function for simplicity, i. e.,

$$\varepsilon(t) = \hat{\varepsilon} \sin 2\pi f t, \tag{6.64}$$

where $\hat{\varepsilon}$ is the amplitude of the strain. The stress response of the fractal dashpot can be written as

$$\sigma = \frac{2\pi f \hat{\varepsilon} \eta}{\alpha} \cdot t^{1-\alpha} \cos 2\pi f t. \tag{6.65}$$

The stress responses of the Abel dashpot and the Newton dashpot are presented as

$$\sigma = \frac{2\pi f \hat{\varepsilon} \xi}{\Gamma(1-\beta)} \int_0^t \frac{\cos 2\pi f \tau}{(t-\tau)^\beta} d\tau \tag{6.66}$$

and

$$\sigma = 2\pi f \hat{\varepsilon} \mu \cdot \cos 2\pi f t. \tag{6.67}$$

For the sake of convenience, all the material parameters and the amplitude and frequency of stress are set as 1, i. e., $\eta = \mu = \xi = f = \hat{\sigma} = 1$. The stress responses of three dashpots are shown in Figure 6.27 and Figure 6.28.

From Figures 6.27 and 6.28, it can be seen that the stress amplitude of the fractal dashpot increases with time and also increases with a smaller value of the derivative order. On the other hand, the stress amplitudes of the Abel dashpot and the Newton dashpot show an independent relation with time and the derivative order. In addition,

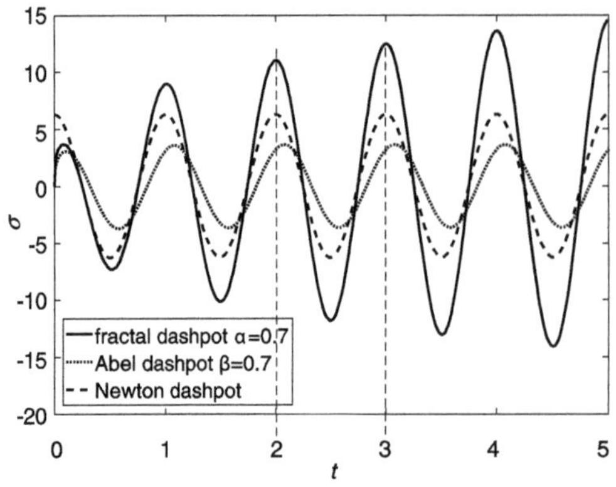

Figure 6.27: The stress responses of the fractal dashpot, Abel dashpot, and Newton dashpot under dynamic strain at derivative order 0.7.

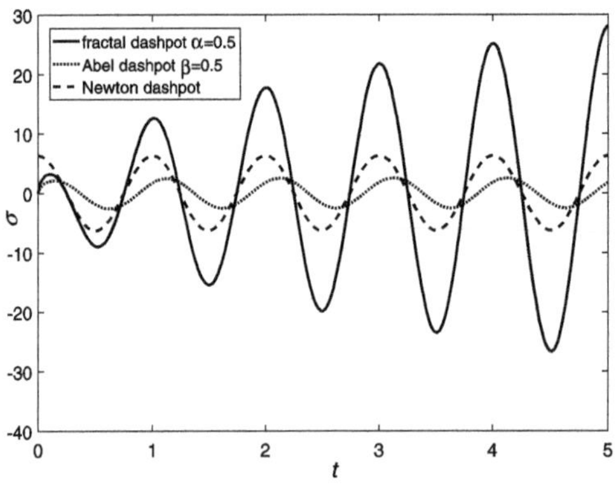

Figure 6.28: The stress responses of the fractal dashpot, Abel dashpot, and Newton dashpot under dynamic strain at derivative order 0.5.

the stress response of the fractal dashpot has the same phase as the Newton dashpot, which is different from that of the Abel dashpot.

In order to extend the use of the fractal dashpot, parallel and in series combinations with spring-pot and Saint-Venant's body can be taken. Based on the idea of the classical component system, these models can be called fractal component models. By taking a combination of the fractal dashpot in parallel with Saint-Venant's body, the fractal Bingham model is introduced. The schematic diagram of the fractal Bingham model is shown in Figure 6.29.

The constitutive relation of the fractal Bingham model can be written as follows:

$$\begin{aligned} &\varepsilon = 0, \quad \sigma < \sigma_s, \\ &\sigma = \sigma_s + \eta \cdot \frac{d\varepsilon}{dt^{\alpha}}, \quad \sigma > \sigma_s, \end{aligned} \tag{6.68}$$

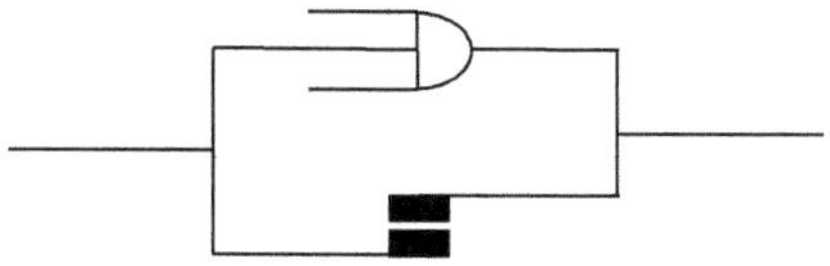

Figure 6.29: Schematic diagram of the fractal Bingham model.

where σ_s represents the yield stress. The stress response of the fractal Bingham model under constant strain rate $\dot{\varepsilon} = c$ is written as

$$\sigma = \sigma_s + \frac{\eta \cdot c \cdot t^{1-\alpha}}{\alpha} \quad \text{or} \quad \sigma = \sigma_s + \frac{\eta \cdot c^{\alpha}}{\alpha} \cdot \varepsilon^{1-\alpha}. \tag{6.69}$$

Under power law strain rate $\dot{\varepsilon} = c \cdot t^n$, the stress response of the fractal Bingham model is presented as

$$\sigma = \sigma_s + \frac{\eta \cdot c}{\alpha} \cdot t^{n+1-\alpha} \quad \text{or} \quad \sigma = \sigma_s + \frac{\eta}{\alpha} \cdot c^{\frac{1}{n}\cdot(\alpha-1)} \cdot \dot{\varepsilon}^{1+\frac{1}{n}\cdot(1-\alpha)}, \tag{6.70}$$

where n is a positive number and $n \geqq 1$.

6.4.2 Applying the fractal dashpot and fractal Maxwell model to creep behaviors

Figures 6.30 and 6.31 show the creep experimental data of polymethyl methacrylate (PMMA) [82] and PMR-15 (one kind of polyimides) [170] as well as the simulation results of the fractal dashpot. Here, the following formula of the relative error (*Rerr*) is introduced to analyze the accuracy of the proposed model:

$$\text{Rerr} = \left|\frac{y - \hat{y}}{y}\right|, \tag{6.71}$$

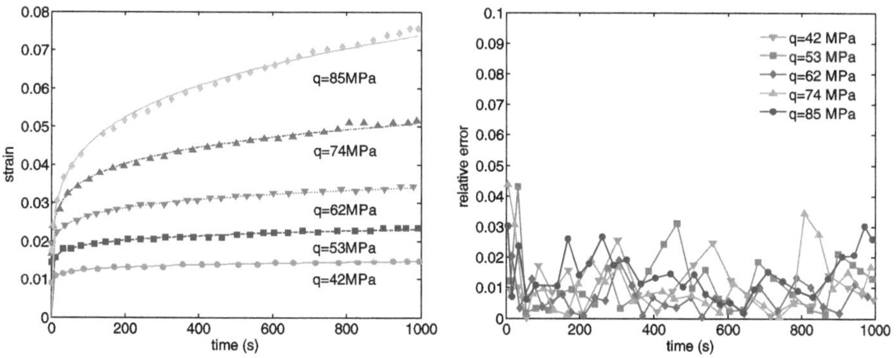

Figure 6.30: The creep experimental data of PMMA and the simulation results of (a) the fractal dashpot and (b) the relative error of strain.

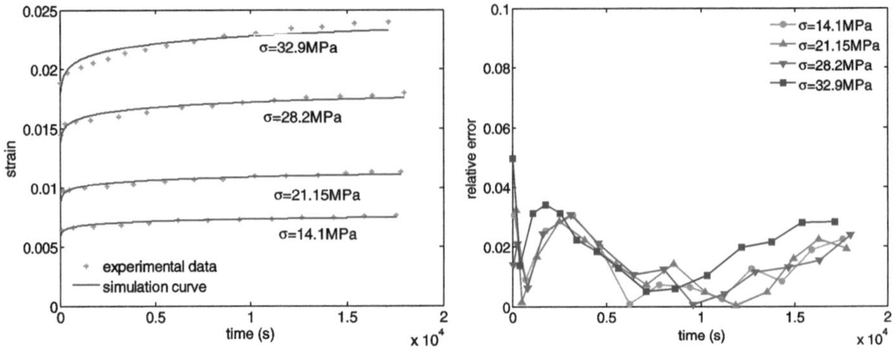

Figure 6.31: The tension creep experimental data of PMR-15 and simulation results of (a) the fractal dashpot and (b) the relative error of strain.

where y and $\hat{y}$ represent the experimental data and the theoretical value, respectively. Figure 6.30(b) and Figure 6.31(b) show the relative errors between the theoretical values and the experimental data. It can be seen that the fractal dashpot is in good agreement with the experimental data. The obtained parameters can be found in Tables 6.10 and 6.11. It can be seen that the higher the stress is, the bigger the derivative order is. In addition, the best-fitting derivative orders are all between 0 and 1, which means PMMA and PMR-15 clearly exhibit viscoelasticity during creep. Furthermore, their viscosities are higher with larger stress.

Figure 6.32(a) shows the creep experimental data of polypropylene copolymer containing short glass fibers (PP-G) [57] and simulation results of the fractal Maxwell

Table 6.10: The parameters from fitting creep experimental data of PMMA by the fractal dashpot.

	η	α
$q = 42$ MPa	4.65e3	0.071
$q = 53$ MPa	4.11e3	0.0847
$q = 62$ MPa	3.82e3	0.1074
$q = 74$ MPa	4.01e3	0.1466
$q = 85$ MPa	5.01e3	0.2128

Table 6.11: The parameters from fitting experimental data of PMR-15 by the fractal dashpot.

	α	η
$\sigma = 14.1$ MPa	0.0354	2.6739e3
$\sigma = 21.15$ MPa	0.0354	2.6906e3
$\sigma = 28.2$ MPa	0.0365	2.2985e3
$\sigma = 32.9$ MPa	0.0416	2.1201e3

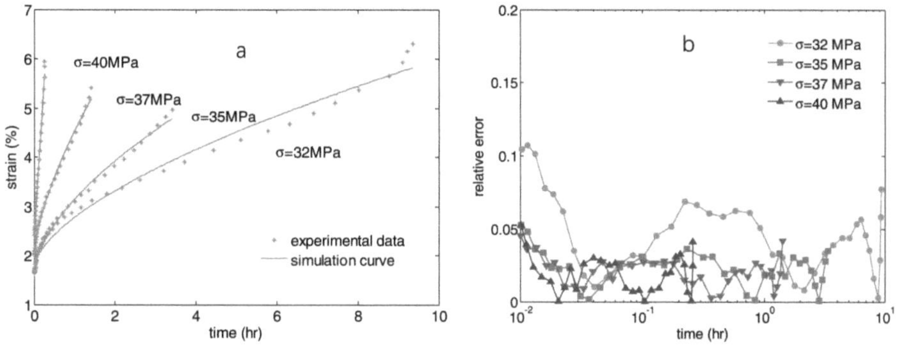

Figure 6.32: The creep experimental data of PP-G and simulation results of (a) the fractal Maxwell model and (b) the relative error of strain.

Table 6.12: The parameters from fitting creep experimental data by the fractal Maxwell model.

	E	η	α
σ = 32 MPa	17.9962	31.9897	0.6249
σ = 35 MPa	21.6354	22.8882	0.5937
σ = 37 MPa	18.8202	14.4057	0.6568
σ = 40 MPa	16.6189	3.1636	0.9948

model. The best-fitting parameters are presented in Table 6.12. In Figure 6.32(b), the fractal Maxwell model shows high accuracy. It can be seen from Table 6.12 that the larger the given stress is, the bigger the derivative order α that can be obtained is, except for the situation $\sigma = 32$ MPa, which means the viscosity of PP-G is larger and increases with increasing stress during creep. It is worth mentioning that PP-G shows an initial elastic phase during creep, which is very suitable to be characterized by the fractal Maxwell model.

6.4.3 Applying the fractal dashpot to describing time-dependent viscosity

Figure 6.33(a) shows the stress response of poly-para-phenylene (PPP) [193] under different constant strain rates at 175°C. It can be seen that stress of PPP increases with time (or strain) at specific strain rates, which reflects the increasing viscosity of PPP during the shear process, that is, rheopexy. Applying the fractal dashpot to fitting the experimental data, the simulation results and parameters can be found in Figure 6.33 and Table 6.13. From Figure 6.33(b) it can be seen that the stress response of the fractal dashpot is in good agreement with experimental data.

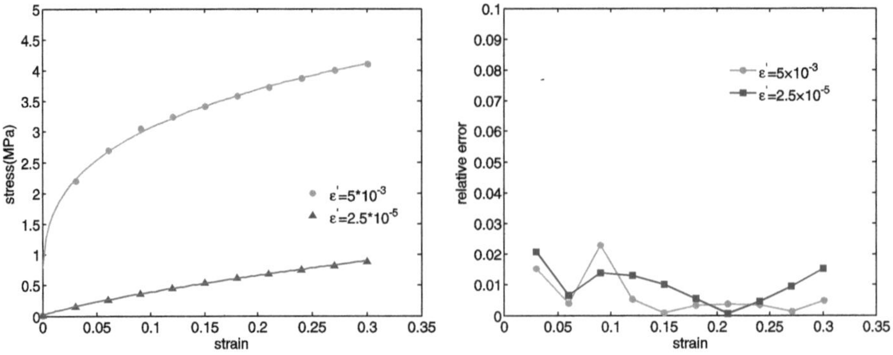

Figure 6.33: The stress response of PPP at constant strain rate and simulation results of (a) the fractal dashpot and (b) the relative error of stress.

Table 6.13: The parameters from fitting constant-strain rate stress response by the fractal dashpot.

	α	η
$\varepsilon' = 5\times10^{-3}$/s	0.7319	200.7924
$\varepsilon' = 2.5\times10^{-5}$/s	0.2406	6.8749

We introduce a concept of the relative strength of rheopexy or thixotropy (r) by the ratio of change rates of apparent viscosity in two situations, i. e.,

$$r = \frac{d\eta_{b1}}{dt} \bigg/ \frac{d\eta_{b2}}{dt} = \frac{\eta_1}{\eta_2} \cdot \frac{1-\alpha_1}{1-\alpha_2} \cdot \frac{\alpha_2}{\alpha_1} \cdot t^{\alpha_2-\alpha_1}. \tag{6.72}$$

If $r > 1$, it means that the rheopexy (or thixotropy) phenomenon in the first situation is more obvious than that in the second situation. If $r < 1$ or $r = 1$, our conclusion is opposite to the one above.

Using equation (6.72) to calculate the relative strength (r) of rheopexy at two situations, $\dot{\varepsilon} = 5\times10^{-3}$ and $\dot{\varepsilon} = 2.5\times10^{-5}$, the result is written as

$$r = 3.3896 \times t^{-0.4913}, \tag{6.73}$$

where r is a monotonic decreasing function of time and $r > 1$ when $t < 11.9971$ s. It means that at the initial time, apparent viscosity increases more quickly under high strain rate. After that time point, apparent viscosity increases more quickly under low strain rate. It can be concluded that at the initial phase of the shear process, a high strain rate is helpful to improve rheopexy of PPP. But rheopexy will become stronger at low strain rate with the passing of time.

Figure 6.34(a) shows the apparent viscosity experimental data [166] of waxy potato starch (WPS) at constant strain rates. Applying equation (6.61) to fitting these experimental data, the simulation results and obtained parameters can be found in

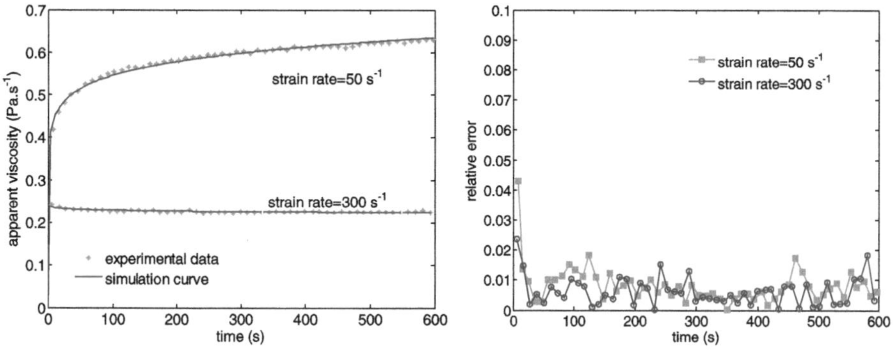

Figure 6.34: The viscosity experimental data of WPS at constant strain rate and simulation results of (a) the fractal dashpot and (b) the relative error of apparent viscosity.

Table 6.14: The parameters from fitting viscosity experimental data by the fractal dashpot.

	α	η
Strain rate = 50 (s^{-1})	0.9149	0.3370
Strain rate = 300 (s^{-1})	1.0116	0.2443

Figure 6.34(a) and Table 6.14. The relative error of fitting curves and experimental data are also presented in Figure 6.34(b). The maximal relative error is less than 5×10^{-2}, which illustrates a high accuracy of the fractal dashpot with experimental data. From Table 6.14 we can see that the derivative order is less than 1 at a strain rate of 50 s^{-1} and is larger than 1 at a strain rate of 300 s^{-1}, which means WPS exhibits rheopexy at small strain rates and thixotropy at large strain rates.

6.4.4 Applying the fractal Bingham model to simulating shear test of muddy clay

Yin et al. [201] conducted the pulling sphere test of muddy clay with different shear rate functions, i. e., $\dot{\varepsilon} = 0.015\,s^{-1}$, $\dot{\varepsilon} = 0.375 \times 10^{-3}\,t\,s^{-1}$, and $\dot{\varepsilon} = 1.215 \times 10^{-4}\,t^2\,s^{-1}$. The experimental data and the simulation results by the fractal Bingham model and the fractional Bingham model are shown in Figure 6.35(a)–(c).

The stress response at constant strain rate, which exhibits time-dependent viscosity, is shown in Figure 6.35(a). From Figures 6.35(b) and (c), it can be seen that the muddy clay behaves as shear thickening fluid with time-linear strain rate and a shear thinning property is seen when the strain rate accelerates. That can also be deduced from the fitting parameters in Table 6.15, where the derivative order is less than 1 at linear strain rate and bigger than 1 at accelerating strain rate. It confirms again that the fractal dashpot can describe the shear thickening property with $0 < \alpha < 1$ and the shear thinning phenomenon with $1 < \alpha < 2$. It is necessary to mention that the

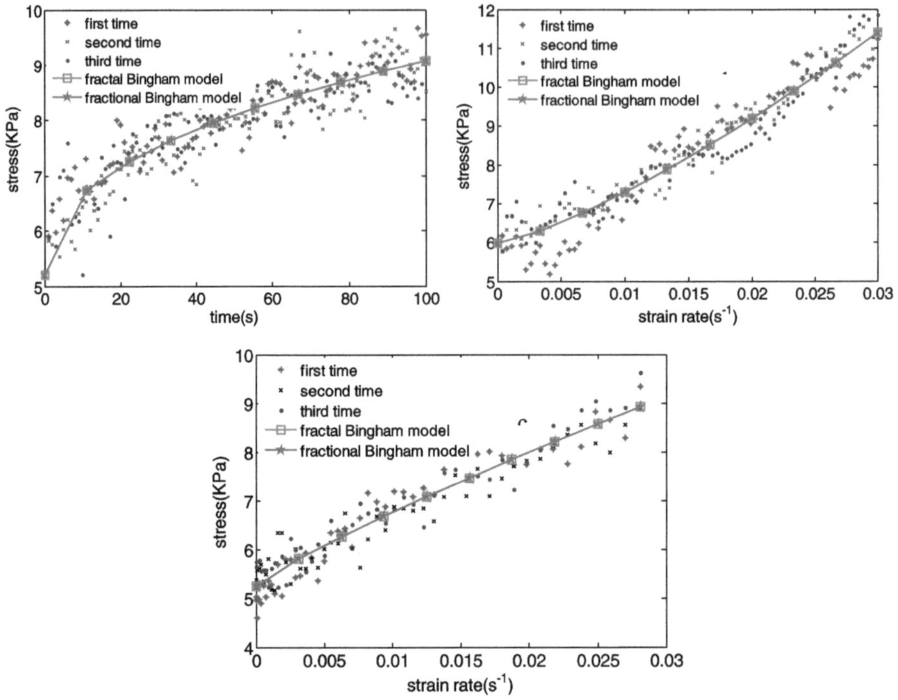

Figure 6.35: (a) The stress-time experimental data for $\dot{\varepsilon} = 0.015\,s^{-1}$, (b) the stress-strain rate experimental data for $\dot{\varepsilon} = 0.375 \times 10^{-3}\,t\,s^{-1}$, and (c) $\dot{\varepsilon} = 1.215 \times 10^{-4}t^2\,s^{-1}$, as well as the simulation results by the fractal Bingham model and the fractional Bingham model.

Table 6.15: The parameters from fitting experimental data by the fractal Bingham model.

	σ_s	η	α
$\dot{\varepsilon} = 0.015\,s^{-1}$	5.2077	21.3312	0.5778
$\dot{\varepsilon} = 0.375 \times 10^{-3}\,t\,s^{-1}$	5.9740	36.3644	0.7116
$\dot{\varepsilon} = 1.215 \times 10^{-4}\,t^2\,s^{-1}$	5.2594	359.2366	1.2813

fractal Bingham model has the same behavior in characterizing the shear-dependent viscosity for non-Newtonian fluid as the fractional Bingham model.

6.4.5 Remarks

As a new viscoelastic component, the fractal dashpot was proposed using the fractal derivative modeling approach. The creep compliance and relaxation modulus of the fractal dashpot were derived. Through a series of loading experiments, the fractal dashpot demonstrated similar viscoelastic characteristics to the Abel dashpot when the derivative order varies from 0 to 1, while the fractal dashpot has a clear advantage of computational efficiency over the Abel dashpot. The Newton dashpot is a special case of the fractal dashpot with derivative order equivalent to 1.

On the other hand, the fractal dashpot was found suitable to characterize the time- and shear-dependent viscosity of non-Newtonian fluid under constant and power law strain rates. As an extension of the fractal dashpot, a fractal Bingham model was also introduced.

The proposed fractal models were validated with high accuracy compared with rheological experimental data of non-Newtonian fluid. The fitted derivative orders increased with increasing stress during creep. This indicates that higher viscosities are achieved under higher stress for these materials. According to the derived relative strength of rheopexy of PPP material, a high strain rate is helpful to improve the strength of rheopexy at the initial phase. With the evolution of time, a low strain rate, however, is observed to enhance rheopexy. In addition, the muddy clay exhibited an obvious plasticity and an increased viscosity over time at constant strain rate. It also behaved as shear thickening fluid at time-linear strain rate and shear thinning fluid at time-accelerated strain rate.

6.5 Richards equation

Based on the work of Sun et al. [177] and Hooshyar and Wang [87], this section uses the Hausdorff Richards equation to derive a generalized water infiltration rate in heterogeneous soil and analyzes its relationship with the existing hydrological models, in which the Hausdorff derivative order is capable of depicting the heterogeneity of the soil.

6.5.1 Hausdorff Richards equation

The Richards equation [196] is often used to describe the vertical motion of water in porous media. From the perspective of diffusion, the Richards equation can be expressed as

$$\frac{\partial \theta}{\partial t} = \frac{\partial}{\partial z}\left(D(\theta)\frac{\partial \theta}{\partial z}\right) - \frac{\partial K(\theta)}{\partial z}, \tag{6.74}$$

where the diffusivity of soil water is defined as

$$D(\theta) = K(\theta)\frac{\partial \Psi_m(\theta)}{\partial \theta}, \tag{6.75}$$

where z is the distance from the datum, the downward direction is positive, θ is the soil moisture content, $K(\theta)$ is the soil conductivity, and $\Psi_m(\theta)$ is the soil matrix potential.

According to the definition of the Hausdorff derivative [31], we can obtain the following time Hausdorff derivative Richards equation [177], which is called the Hausdorff Richards equation:

$$\frac{\partial\theta}{\partial t^\alpha} = \frac{\partial}{\partial z}\left(D(\theta)\frac{\partial\theta}{\partial z}\right) - \frac{\partial K(\theta)}{\partial z}, \tag{6.76}$$

where $\frac{\partial\theta}{\partial t^\alpha} = \lim_{\Delta t\to 0}\frac{\theta(t+\Delta t)-\theta(t)}{(t+\Delta t)^\alpha - t^\alpha}$ and $0 < \alpha \le 1$ is the order of the time Hausdorff derivative, which can describe the Hausdorff dimension of the particle diffusion trajectory. The dimension of the soil water diffusion rate is m^2/s^α, and the dimensions of other variables are the same as for the classical Richards equation. The Hausdorff Richards equation is capable of characterizing the non-Boltzmann scale in the process of soil water infiltration. When $\alpha = 1$, it reduces to the classical Richards equation. When $\alpha \ne 1$, the Hausdorff Richards equation can depict the non-exponential attenuation of soil moisture content with time. In other words, it is slower than the classical Richards equation corresponding to the exponential decay.

It can be noted by equation (6.76) that the capillary force and gravity jointly determine the infiltration process. In the early stage of the infiltration process, the capillary force plays a key role and the gravity can be ignored. The corresponding control equation is

$$\frac{\partial\theta}{\partial t^\alpha} = \frac{\partial}{\partial z}\left(D(\theta)\frac{\partial\theta}{\partial z}\right). \tag{6.77}$$

When the diffusivity is constant, i. e., $D(\theta) = D$, equation (6.77) reduces to

$$\frac{\partial\theta}{\partial t^\alpha} = D\frac{\partial^2\theta}{\partial z^2}. \tag{6.78}$$

According to the different initial values and boundary conditions, the relationship between soil infiltration rate and time in different forms can be deduced to

$$f = -D\frac{\partial\theta}{\partial z}(z = 0, t), \tag{6.79}$$

in which the initial value of the soil infiltration rate is $f(t = 0) = f_0$.

6.5.2 Soil infiltration rate

In order to facilitate the derivation of the infiltration rate, another equivalent form of equation (6.78) is adopted [87], i. e.,

$$\frac{\partial\tau}{\partial t^\alpha} = D\frac{\partial^2\tau}{\partial z^2}, \tag{6.80}$$

where τ is the relative water shortage of the soil and the relationship between τ and soil moisture content θ is

$$\tau = 1 - \frac{\theta - \theta_r}{\theta_s - \theta_r}, \tag{6.81}$$

where θ_r and θ_s are residual moisture content and saturated moisture content, respectively.

Based on the initial boundary value in [87], the soil infiltration rate was deduced as

$$\tau(z = l, t = 0) = \tau_0, \tag{6.82}$$

$$\tau(z = 0, t) = 0, \tag{6.83}$$

$$\frac{\partial \tau}{\partial z}(z = l, t) = 0, \tag{6.84}$$

where $l = l_0 \tau^{-1/\lambda}$ and λ is the aperture distribution index.

According to the classical separation variable method, the solution of equation (6.80) can be obtained as

$$\tau = \left(\frac{\pi^2 D}{2\lambda l_0^2} t^\alpha + 1 \right)^{-\lambda/2} \left(\frac{\theta_s - \theta_0}{\theta_s - \theta_r} \right) \sin\left(\frac{\pi}{2l(t)} z \right). \tag{6.85}$$

Combining equation (6.79) and equation (6.81), the soil infiltration rate is

$$f = \frac{D\pi(\theta_s - \theta_0)}{2l_0} \left(\frac{\pi^2 D}{2\lambda l_0^2} t^\alpha + 1 \right)^{-\lambda/2-1/2}. \tag{6.86}$$

According to the initial value of the infiltration rate, equation (6.86) can be simplified as

$$f = f_0 \left(\frac{\pi f_0}{\lambda l_0 (\theta_s - \theta_0)} t^\alpha + 1 \right)^{-\lambda/2-1/2}. \tag{6.87}$$

When $\alpha = 1$, the corresponding results of equation (6.87) are consistent with the results given by Hooshyar and Wang [87]. The value of λ determines the infiltration rate of the corresponding hydrological model.

For instance, when $\lambda = 3$, equation (6.87) is the generalization of the model of the runoff curve (SCS-CN) [140], abbreviated as SCS-CN model [43], i. e.,

$$f = f_0 \left(\frac{f_0}{S} t^\alpha + 1 \right)^{-2}, \tag{6.88}$$

where S is the potential soil water retention rate.

6.5.3 Analysis of examples

In order to investigate the influence of the order of the time Hausdorff derivative on the soil infiltration rate, Figure 6.36 shows the curve of the infiltration rate for four different values of α, where $\lambda = 3$, $D = 1$, $l_0 = 1$, and $\theta_s - \theta_0 = 1$. The influence of different values of λ on the soil infiltration rate is also investigated. According to equation (6.86), Figure 6.37 shows the variation curve of the infiltration rate in terms of time for four different values of λ, in which $\alpha = 0.6$, $D = 1$, $l_0 = 1$, and $\theta_s - \theta_0 = 1$.

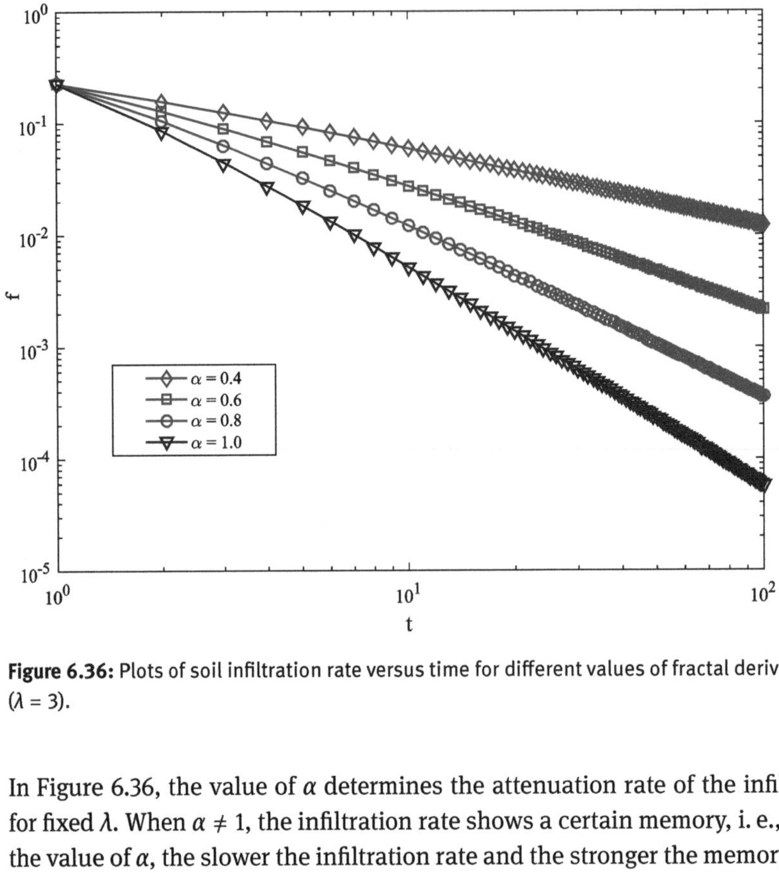

Figure 6.36: Plots of soil infiltration rate versus time for different values of fractal derivative order ($\lambda = 3$).

In Figure 6.36, the value of α determines the attenuation rate of the infiltration rate for fixed λ. When $\alpha \neq 1$, the infiltration rate shows a certain memory, i. e., the smaller the value of α, the slower the infiltration rate and the stronger the memory. Moreover, it can reflect the heterogeneous diffusion environment of water infiltration, which deviates much from the hypothesis of uniform medium from classical models. On the one hand, the memory can be interpreted such that the Hausdorff derivative value is not equal to 1, and the infiltration rate is slower than the classical Richards model of infiltration rate attenuation. On the other hand, the Hausdorff Richards model can describe a class of heterogeneous media by abnormal infiltration processes. Therefore, the order of the Hausdorff derivative is an important parameter to characterize

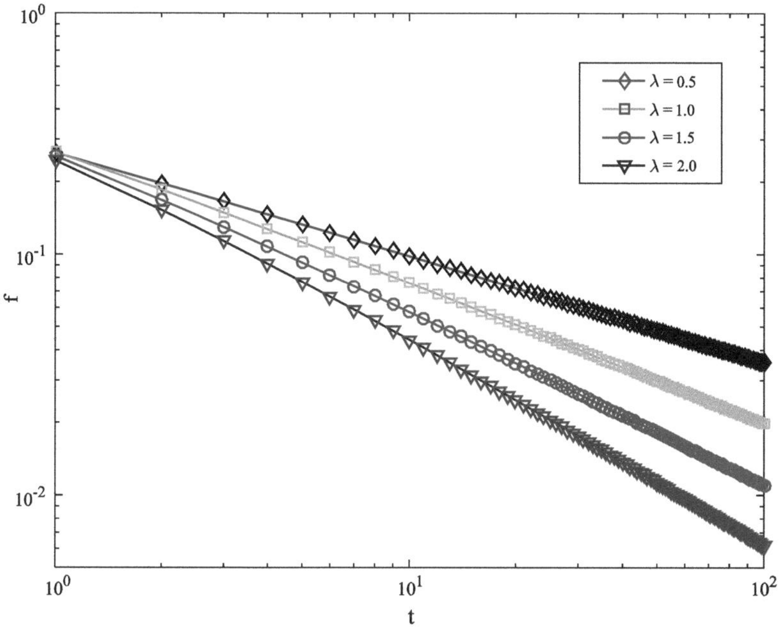

Figure 6.37: Plots of soil infiltration rate versus time for different values of the pore size distribution index (α = 0.6).

soil structure, which can reflect the heterogeneity of soil. As illustrated in Figure 6.37, with fixed α, λ determines the attenuation rate of the infiltration rate. The smaller the values of λ are, the slower the soil moisture permeates. Thus, the soil microstructure is another important index.

Based on the above Hausdorff Richards equation, the relationship between soil infiltration rate and time is derived. The Hausdorff derivative order can characterize the complexity of the water diffusion environment in soil and the heterogeneity structure of soil. From the results of the two examples, we can observe that when $\alpha \neq 1$, the soil infiltration rate shows a certain memory, i. e., the smaller the value of α, the slower the infiltration rate and the stronger the memory. It also reflects the complexity of the diffusion environment of water infiltration. In addition, the smaller the values of λ is, the smaller the soil moisture permeability is. This is another important indicator to characterize the soil structure.

It is also noted that the relationship between soil infiltration rate and time can be generalized from the perspective of the fractional Darcy law and the fractional derivative Richards equation. Combining with the specific experimental data, the relationship between the Hausdorff derivative order and the distribution of soil pore size as well as the difference and connection of the existing hydrological models will be considered in the next step.

6.6 Magnetic resonance imaging

Diffusion-weighted magnetic resonance imaging (DW-MRI) is a quantitative method that can directly probe the movement of water protons in biological tissues [2]. It is known that the DW-MRI decay signal for water diffusion follows the exponential form of the Stejskal–Tanner equation in a pulse-gradient spin-echo experiment for unrestricted, classical Brownian motion [25]. However, due to trapping, tortuosity, and turbulence in heterogeneous biological tissues such as brain white and gray matter, the diffusion decay signal deviates from exponential decay when water movement is restricted by obstacles persisting over different lengths and time scales [146]. In this section, we adopt the fractal derivative diffusion model to describe water diffusion in MRI, and we apply the model in the examination of fixed mouse brain tissue with the ultimate goal to differentiate between white and gray matter based on their characteristic sub-voxel tissue microstructure.

6.6.1 The fractal derivative model for MRI signals

In pulse-gradient spin-echo experiments, the MRI signal attenuation is usually assumed to obey the Stejskal–Tanner decay equation with a mono-exponential form, which is expressed as

$$S/S_0 = \exp(-bD), \tag{6.89}$$

where S is the signal magnitude, S_0 is the signal magnitude at lowest b value, D is the apparent diffusion coefficient (ADC) ($\mathrm{m^2\,s^{-1}}$), $b = q^2\bar{\Delta}$, q is the diffusion gradient strength sensitization with units of radians/m, and $q = \gamma G\delta$, $\bar{\Delta} = \Delta - \delta/3$, where γ is the gyromagnetic ratio, G is the amplitude of the diffusion gradient (Tesla/m), Δ is the observation time between gradient pulses (ms), and δ is the pulse length (ms) [126].

The following expression is used to quantify the signal intensity decay in DW-MRI [115]:

$$p(q,\bar{\Delta}) = \exp(-D_{\alpha,\beta}q^{2\beta}\bar{\Delta}^{\alpha}), \tag{6.90}$$

which simplifies into equation (6.89) when $\alpha, \beta = 1$.

6.6.2 Spectral entropy of fractal derivative model

Since the solution of the fractal derivative model has explicit form in terms of the spatial frequency, q, we can describe the diffusion processes in "q-space" by using the spectral entropy [127]. Thus, for varying Δ and fixed q, the normalized spectral entropy is stated as

$$H = -\sum_{i=1}^{N} \frac{\hat{p}(q,\bar{\Delta}_i) ln(\hat{p}(q,\bar{\Delta}_i))}{ln(N)}, \tag{6.91}$$

where

$$\hat{p}(q,\bar{\Delta}_i) = \frac{p(q,\bar{\Delta}_i)p^*(q,\bar{\Delta}_i)}{\sum_{i=1}^{N}(p(q,\bar{\Delta}_i)p^*(q,\bar{\Delta}_i))}, \tag{6.92}$$

$p^*(q,\bar{\Delta})$ is the complex conjugate of $p(q,\bar{\Delta})$, and N is the total number of observation times. The normalized power density, $\hat{p}(q,\bar{\Delta}_i)$, describes the spatial correlation, and its sum is equal to one. Equation (6.91) is adapted from the Shannon information entropy and the spectral entropy given by Viertio-Oja et al. [183]. The range of spectral entropy H is between 0 and 1.

The spectral entropy of a diffusion process in biological tissue can be measured using the fractal derivative equation by inserting equation (6.90) into equation (6.91). Additional details about the use of spectral entropy to describe anomalous diffusion processes in DW-MRI can be found in [91, 92].

6.6.3 MR system and diffusion sequences

The DW-MRI experiments were conducted at the Advanced Magnetic Resonance Imaging and Spectroscopy (AMRIS) facility, in Gainesville, FL. The observation time was varied from 15.6 to 115 ms, while the pulse length $\delta = 3.5$ ms. The b values for the experiments ranged from 121.5 to 22,700 s/mm^2. The amplitude of the diffusion encoding gradient was $G = 400$ mT/m. The additional imaging parameters are $100 \times 100 \times 500\ \mu m^3$ voxel resolution, echo time TE = 22.9 ms, and repetition time TR = 3000 ms. More details are given in [200].

The attenuation signals are initially normalized by dividing by S_0, which is the signal magnitude at $b = 0$. The fractal derivative diffusion equation (6.90) is used to fit the attenuation signals pixel by pixel. In equation (6.90), we set $\beta = 1$ because we expect the *ex vivo* tissue to show sub-diffusive properties and greater sensitivity in the parameter α when varying the observation time Δ. The remaining two parameters α and $D_{\alpha,\beta}$ in equation (6.90) were determined for all the pixels by estimating the values using a non-linear least squares regression.

6.6.4 Results

Whole brain analysis was performed by applying the fractal derivative model to MR data collected from each slice of the fixed mouse brain. Figure 6.38 shows the regions

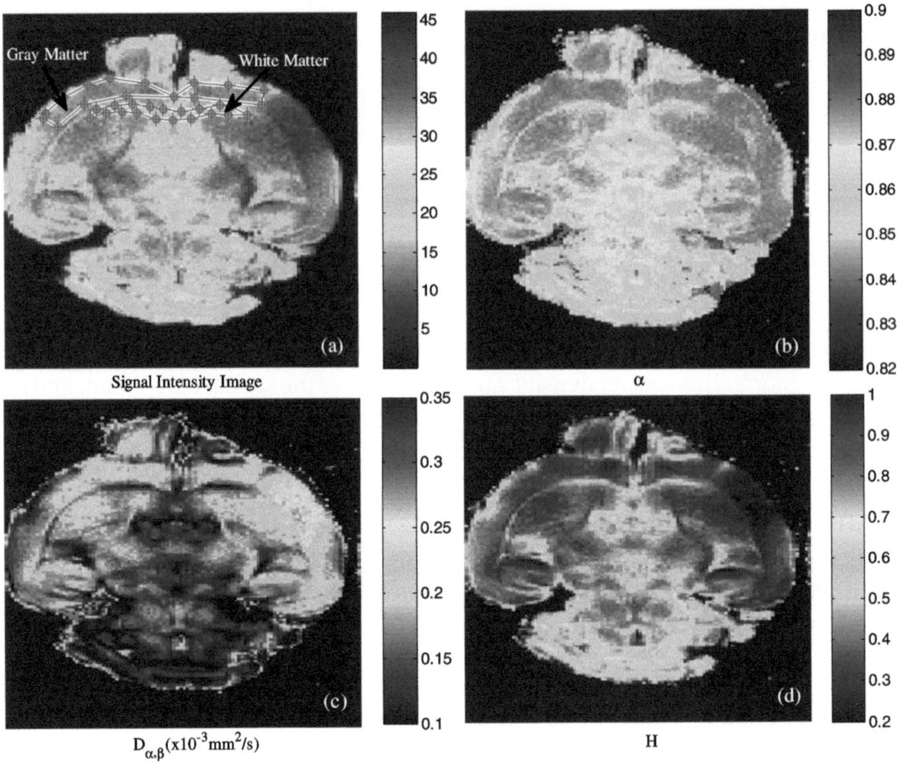

Figure 6.38: Axial slice through the white matter and gray matter regions of a mouse brain with the corresponding maps of (a) MR signal intensity with selected ROIs in white and gray matter, (b) the order of the time fractal derivative, (c) the diffusion coefficient, and (d) the spectral entropy.

of interest (ROIs) in the white matter and gray matter and the parameter maps for the fractal derivative model in equation (6.90) and spectral entropy in equation (6.91). The contrast in Figure 6.38 shows that the α parameter clearly separates white from gray matter. In addition, the white matter appears more heterogeneous and tortuous compared to the gray matter. In the $D_{\alpha,\beta}$ map, the contrast between white matter and gray matter is clearly visible, but the white and gray matter tissue regions appear less heterogeneous than in the α parameter map. The spectral entropy H map was able to distinguish white matter from gray matter and to provide contrast within the white and gray tissue regions. It was also less noisy than the individual parameter maps.

Given the well-known histological differences between white and gray matter in the brain [149] (e. g., white matter consists of multiscale bundles of self-similar, anisotropic axons, fibers, and fiber tracts, while gray matter consists of a macroscopically layered, but microscopically random network of neuron and axons), it is reasonable to assume that the spectral entropy, H, conveys more details about the microstructure of the white matter than the gray matter, thus H appears to be a more effective parameter for displaying specific brain tissue structures than α or $D_{\alpha,\beta}$. Overall, voxels exhibiting more tissue complexity should exhibit a lower α, a lower diffusion coefficient, and a higher spectral entropy, reflecting (i) bounded diffusion and propagation of water in regions with greater local interactions or (ii) restricted excursion of water molecules in the microstructure of white matter compared to gray matter areas of the mouse brain. One way of illustrating this behavior is to plot histograms of all the corresponding parameter values for both tissues.

Figure 6.39 gives the distributions of α, $D_{\alpha,\beta}$, and H for ROIs selected in the white matter and gray matter as shown in Figure 6.38. Table 6.16 reports the corresponding mean and standard deviation of α, $D_{\alpha,\beta}$, and H. In Figure 6.39, all the distributions are assumed to be Gaussian distributions, and the mean and variance determine its shape. We can see from Figure 6.39 that the distributions clearly differentiate between gray matter and white matter, particularly in the distribution density of α. Thus, the distribution density can be considered as a potential measure with which to separate or classify biological tissues. On the other hand, the mean values for the order of the fractal derivative (α) and the derived fractal dimension (F) in the white matter and gray matter are very similar. Thus, while we can use α and F to interpret the characteristics of the diffusion trajectory, for selected ROIs, we find the values of $D_{\alpha,\beta}$ and H to be more useful measures of the tortuosity of white and gray matter.

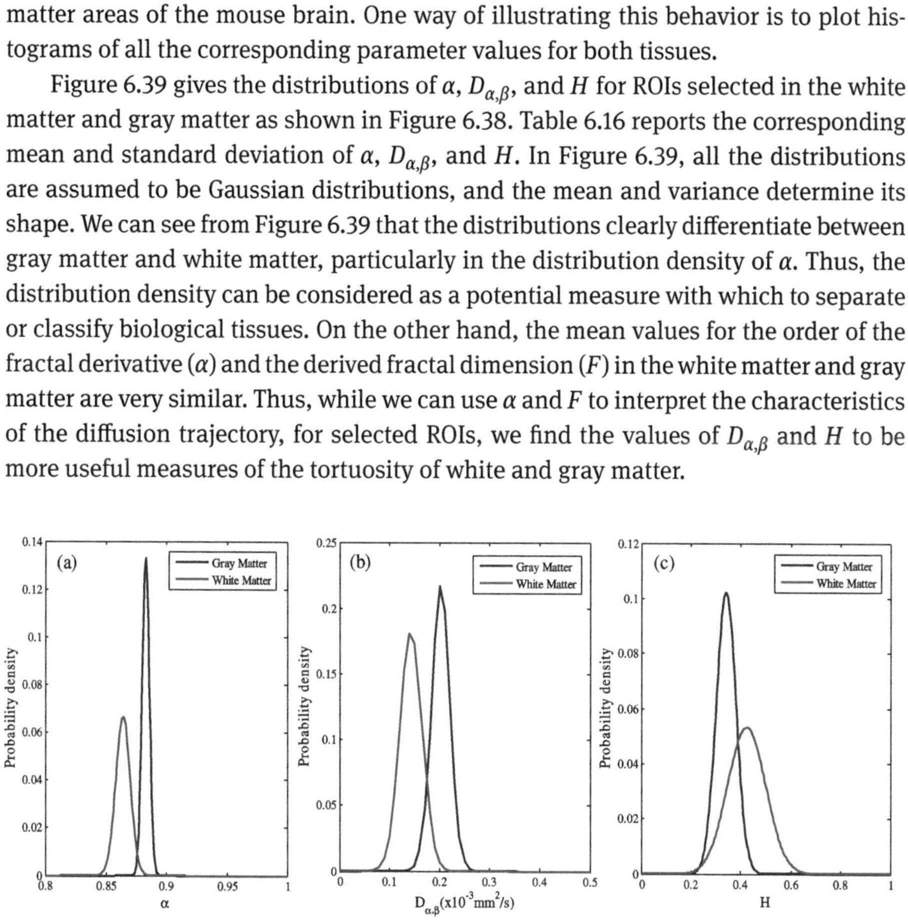

Figure 6.39: Probability density distributions for the values of (a) time fractal derivative order α, (b) diffusion coefficient $D_{\alpha,\beta}$, and (c) spectral entropy H for selected ROIs in the gray and the white matter of an axial brain slice (in each figure the gray matter peak is higher).

Table 6.16: Diffusion and complexity measures for ROIs selected in white matter and gray matter.

ROI	α	$D_{\alpha,\beta}$	H
White matter	0.86±0.01	0.14±0.02	0.42±0.08
Gray matter	0.88±0.01	0.20±0.02	0.34±0.04

Finally, in order to compare and contrast the utility of the fractal derivative model for the analysis of DW-MRI data, we give example fits to the diffusion attenuation signals for the selected ROI used in Figure 6.38 for white and gray matter plotted on a logarithmic-linear graph of signal decay versus *b* value. In Figure 6.40 we compare the experimental data with fits using the mono-exponential model, the stretched exponential model [16], the fractal derivative model, the fractional derivative model [128], and the diffusion kurtosis imaging (DKI) model [94]. From these plots, it is easily seen that with increasing *b* value, the curves estimated by the mono-exponential model and the stretched exponential model decay much faster than the data points and have a lower accuracy compared to the curves estimated using the fractal derivative model. In addition, for the relatively high *b* value data points, the DKI fits poorly to the monotonically decreased signals with increased diffusion weighting, largely due to its parabolic-form fitting function [93]. Compared to the fractional derivative model, the fractal derivative model gives less accurate fitting results for the high *b* values up to 22,700 s/mm^2.

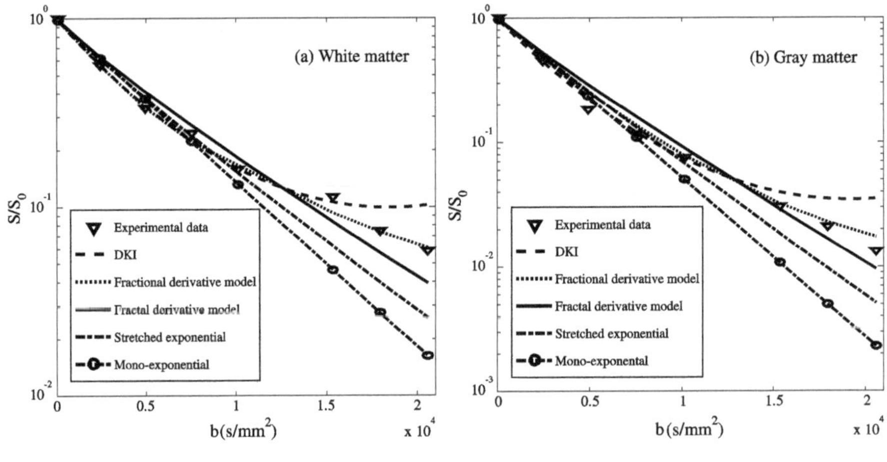

Figure 6.40: Logarithmic plots of the experimental signal decay in (a) the white matter and (b) the gray matter with corresponding fits obtained using five different theoretical models.

To quantitatively compare the fitting accuracy of all the models for the white and gray matter, Tables 6.17 and 6.18 give the slopes of each curve on a logarithmic-linear scale with *b* values smaller and larger than 10,000 s/mm^2, respectively. Considering the results presented in Tables 6.17 and 6.18, for the range of *b* values, all the slopes are consistent with those of the experimental data up to *b* values of 10,000 s/mm^2, while slopes for the fractional-order derivative model are the closest to those of the experimental data with *b* values up to 22,700 s/mm^2. Tables 6.19 and 6.20 provide the corresponding root mean square error (RMSE) of the fitting results for the five models. In addition, the CPU times (RAM 8.0 Gigabytes and Intel Core 3.3 GHz) for the fractal

Table 6.17: Calculated slopes for the five signal decay curves in Figure 6.40(a) plotted on a logarithmic-linear scale for white matter.

	Experimental	DKI	Fractional	Fractal	Stretched	Mono-
$b < 10^4$	−0.19	−0.20	−0.19	−0.18	−0.19	−0.20
$b \geq 10^4$	−0.10	−0.04	−0.10	−0.14	−0.17	−0.20

Table 6.18: Calculated slopes for the five signal decay curves in Figure 6.40(b) plotted on a logarithmic-linear scale for gray matter.

	Experimental	DKI	Fractional	Fractal	Stretched	Mono-
$b < 10^4$	−0.29	−0.28	−0.26	−0.25	−0.27	−0.30
$b \geq 10^4$	−0.17	−0.07	−0.15	−0.21	−0.25	−0.30

Table 6.19: Calculated root mean square error for the five signal decay curves in Figure 6.40(a) for white matter.

	DKI	Fractional	Fractal	Stretched	Mono-
$b < 10^4$	0.019	0.019	0.043	0.034	0.037
$b \geq 10^4$	0.026	0.011	0.021	0.036	0.047

Table 6.20: Calculated root mean square error for the five signal decay curves in Figure 6.40(b) for gray matter.

	DKI	Fractional	Fractal	Stretched	Mono-
$b < 10^4$	0.028	0.043	0.056	0.048	0.035
$b \geq 10^4$	0.014	0.003	0.007	0.010	0.019

and fractional derivative models to analyze the entire images are 10.8 hrs and 13 hrs, respectively. Overall, the fractal model appears to be the second best model over the complete range of b values studied.

6.6.5 Remarks

This section demonstrates that the fractal derivative model can be used to describe the anomalous diffusion process as captured by DW-MRI in a fixed mouse brain. In the model, the fractal dimension of the diffusion trajectory is directly expressed in terms of the Hausdorff fractal dimension, and the spectral entropy is used to measure the uncertainty in the heterogeneous and multiscale system of biological tissues. Interestingly, the fractal derivative order, the diffusion coefficient, the spectral entropy, and their corresponding distribution densities can well differentiate between the white matter and gray matter. Though the results reflect the usual practice, the placements of the ROIs are based on a non-automatic, manual segmentation method. In future work, a semiautomatic or fully automatic segmentation method could be used to further test the proposed fractal derivative model.

In the fractal derivative model we set $\beta = 1$, as we anticipated that the *ex vivo* tissue would exhibit only sub-diffusive properties. In this situation, we observed a greater sensitivity in the parameter α with increasing observation time Δ. Furthermore, in our future work, a more general model fitting both α and β might be considered by varying both observation time and gradient strength in a living animal. As an example, it is intuitive to think about what features of the fractal derivative model would be most useful to detect and characterize *in vivo* changes in experimental animal models of neurological diseases [73]. In such experiments, using a combination of transgenic animals models, MRI and ultrastructural techniques we may be able to correlate changes in the parameters with particular alterations of gray and white matter during the progressive stages of the disease.

Considering potential clinical applications, it is also important to note the results of other diffusion studies. For example, Toivonen et al. [179] were not able to detect or classify prostate cancer using the biexponential, DKI, or stretched exponential models for b values up to 2000 s/mm^2, while Han et al. [78] did observe lesions of medulloblastoma from changes in the diffusion coefficient for b values larger than 1000 s/mm^2, where a deviation from the mono-exponential model appeared. Such changes in the properties of tumors at high b values indicate a potential role for the non-Gaussian models in tumor classification, tissue staging, and disease monitoring. Applying the fractal derivative model, which generalizes the exponential and stretched exponential models, provides a flexible way to fit the DW-MRI decay signal. An additional factor that would impact the clinical utility of the various models is the difficulty of satisfying the short pulse approximation [173] inherent in the fractional-order derivative model of Magin et al. [128] and Carson et al. [91]. Thus, the effects of gradient pulse length as characterized by Gao et al. [70] on a clinical 3T MRI and the recent full fractional-order derivative gradient pulse solutions to the Bloch–Torrey equation by Lin [119] also need to be considered and applied in future clinical studies.

7 Local structural derivative

7.1 General non-Euclidean distance

The Hausdorff distance in space is described by

$$M = |x^d - x'^d|, \tag{7.1}$$

where d is the fractal dimension. The power law in equation (7.1) has since found many applications in science and engineering. However, it is observed that many real problems cannot be well characterized by the power law fractal metrics. Here, the following form generalizes the fractal metric in equation (7.1):

$$M = |R(x) - R(x')|, \tag{7.2}$$

where R is the structural function and can be an arbitrary function. Formula (7.2) characterizes the structural metrics [37], which includes the Euclidean and the non-Euclidean metrics.

7.2 Local structural derivatives

To illustrate the calculus on the structural metrics, we consider the displacement x under power function time metric

$$x = vt^{\alpha}, \tag{7.3}$$

where α is the index of the time fractal.

The corresponding derivative of the displacement in equation (7.3) is obtained as

$$dx = vd(t^{\alpha}). \tag{7.4}$$

Then equation (7.4) can be rewritten as

$$v = \frac{dx}{d(t^{\alpha})}. \tag{7.5}$$

Consequently, the Hausdorff derivative on time fractal α is given by [31]

$$\frac{dp(x,t)}{dt} = \lim_{t_1 \to t} \frac{p(x,t_1) - p(x,t)}{t_1^{\alpha} - t^{\alpha}}. \tag{7.6}$$

By using a similar strategy, the displacement x under the general metric can be derived by

$$x = vk(t), \tag{7.7}$$

where $k(t)$ is the structural function.

https://doi.org/10.1515/9783110608526-007

The derivative of the displacement in equation (7.7) is written as

$$dx = vdk(t). \tag{7.8}$$

Then equation (7.8) can be rewritten as

$$v = \frac{dx}{dk(t)}. \tag{7.9}$$

The definition of structural derivative by using the structural metric further extends the Hausdorff derivative. On the time structural metric, the local structural derivative is defined as [40]

$$\frac{dp(x,t)}{d_s t} = \lim_{t_1 \to t} \frac{p(x,t_1) - p(x,t)}{k(t_1) - k(t)}, \tag{7.10}$$

where S denotes the structural derivative, $k(t)$ is the structural function, and d_s means the structural derivative. When $k(t) = t^\alpha$, it is the Hausdorff derivative. In contrast, the global structural derivative in time is defined as [39]

$$\frac{\delta p(x,t)}{\delta_s t} = \frac{\partial}{\partial t} \int_{t_1}^{t} k(t-\tau) p(x,\tau) d\tau. \tag{7.11}$$

When $k(t) = \frac{t^{-\alpha}}{\Gamma(1-\alpha)}$, equation (7.11) degenerates into the classical Riemann–Liouville fractional derivative.

It is stressed that the structural function is not necessarily a power function. Instead it can be any form of a function, such as the inverse Mittag-Leffler (ML) function [41], the probability density function, and the stretched exponential function. Compared with the classical non-linear calculus models, the structural derivative requires fewer parameters and lower computational costs in detecting the causal relationship between mesoscopic time-space structure and certain physical behaviors [41]. Figure 7.1 shows the decay of four different structural functions, i. e., the exponential, power law, logarithmic, and inverse ML functions.

Overall, the structural function determines the properties and physical mechanisms of the structural derivative and reflects the correlation or mesostructure features of the system. For instance, the mesoscopic interaction causes the diffusion to occur far more slowly. In addition, the structural derivative extends the definitions of the local and non-local derivatives and is more flexible in modeling complex physical phenomena [41].

7.3 Structural derivative models for ultraslow diffusion

Anomalous diffusion including sub-diffusion and super-diffusion has long been an active topic in diverse fields, such as magnetic resonance imaging [91], prime number

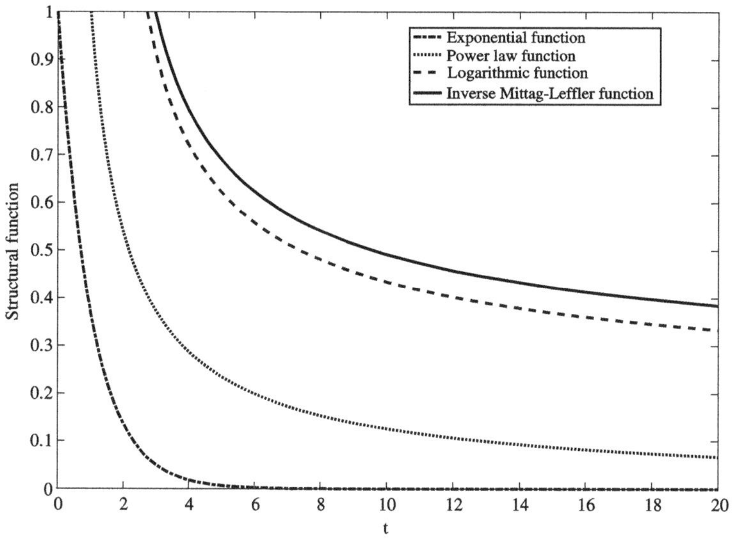

Figure 7.1: Decay of four different structural functions, i. e., the exponential, power law, logarithmic, and inverse ML functions, from bottom to top.

distribution [42], hydrology [160], viscoelastic soft materials [129], and electrochemistry [58], just to mention a few. A variety of models have been proposed to describe the anomalous diffusion, and the mean squared displacement (MSD) provides a common denominator for all these models [134]. Among them, the most popular statistical diffusion model is a power law function of time, i. e.,

$$\langle x^2 \rangle \sim t^{\gamma}, \tag{7.12}$$

where $\langle x^2 \rangle$ represents the MSD. When $\gamma = 1$, equation (7.12) describes the classical Brownian motion, while when $\gamma < 1$, the model is classified as sub-diffusion; and if $\gamma > 1$, it characterizes so-called super-diffusion.

Compared to the anomalous diffusion characterized by the power law function of time, the logarithmic diffusion model [158, 167, 172] has attracted much less attention. It can be stated as

$$\langle x^2 \rangle \sim (\ln t)^{\lambda}, \tag{7.13}$$

where $\lambda > 0$. The statistical model (7.13) describes so-called ultraslow diffusion, which behaves dramatically differently from anomalous diffusion and is even much slower than the so-called sub-diffusion. When $\lambda = 4$, equation (7.13) degenerates into the classical Sinai diffusion law [167], and when $\lambda = 0.5$, it correlates with the well-known Harris law [120].

The logarithmic time evolution has been observed in numerous experiments, for instance, relaxation in Andersson insulators [182], DNA local structure relaxation [20], and paper crumpling under a heavy piston [133]. However, the physical mechanism underlying the logarithmic diffusion model is still missing. In addition, the model cannot describe even slower diffusion processes.

In recent decades, the fractional derivative has become a very popular mathematical approach to develop a variety of partial differential equation models for anomalous diffusion [205]. But this modeling strategy simply does not work for ultraslow diffusion. To remedy this challenging problem, this study presents the structural derivative approach to establish a generalized model for ultraslow diffusion. The concept of the structural function plays a central role in this new strategy as a kernel transform of underlying time-space fabric of physical systems. The existing derivative modeling approaches simply describe the change rate of a certain physical variable with respect to time or space and consider to a lower extent the influence of mesoscopic time-space fabric of a complex system on its physical behavior. Instead, the structural derivative is proposed to describe causal relationships of mesoscopic time-space structures and certain physical behavior in a simple fashion. It is worth noting that the structural derivative can include either a local fractal derivative or a global fractional derivative [41], in which the power function is the structural function. The implicit calculus modeling [34] can be used to construct the structural derivative. The Caputo–Fabrizio fractional derivative [26] and the recent fundamental solution definition of fractional Laplacian [44] are special cases of the proposed structural derivative.

7.3.1 Time structural derivative model for ultraslow diffusion

In the structural derivative modeling of ultraslow diffusion, the propagator $p(x,t)$, i. e., the probability of finding a diffusion particle at position x at time t, is the solution to the diffusion equation

$$\frac{Sp(x,t)}{St} = D_\alpha \frac{\partial^2 p(x,t)}{\partial x^2}, \quad t > 0,\ -\infty < x < \infty, \tag{7.14}$$

where D_α is the generalized diffusion coefficient (m^2/s^α), $p(x,0) = \delta(x)$ represents the assumed initial particle distribution, and the definition of structural derivative in time is given by

$$\frac{Sp(x,t)}{St} = \lim_{t_1 \to t} \frac{p(x,t_1) - p(x,t)}{E_\alpha^{-1}(t_1) - E_\alpha^{-1}(t)}, \tag{7.15}$$

in which $E_\alpha^{-1}(t)$ is the inverse function of the single-parameter ML function [86]. The ML function has the power series form $E_\alpha(t) = \sum_{k=0}^{\infty} \frac{t^k}{\Gamma(1+\alpha k)}$, where Γ denotes the Gamma function, and when $\alpha = 1$, $E_1(t)$ is reduced to the classical exponential function [79]. It is easy to numerically calculate $E_\alpha^{-1}(t)$ by the ML function, and when $\alpha = 1$, $E_1^{-1}(t)$ is reduced to the logarithmic function.

Equation (7.14) can be restated as a normal diffusion equation by using the following kernel transform:

$$\hat{t} = E_\alpha^{-1}(t). \tag{7.16}$$

Then solving equation (7.14) yields a Gaussian distribution, i. e.,

$$p(x,\hat{t}) = \frac{1}{\sqrt{4\pi D_\alpha \hat{t}}} \exp\left(-\frac{x^2}{4D_\alpha \hat{t}}\right). \tag{7.17}$$

The characteristic function of $p(x,t)$, i. e., the Fourier transform of equation (7.17), is given by

$$p(q,t) = \exp(-D_\alpha q^2 E_\alpha^{-1}(t)), \tag{7.18}$$

which describes the relaxation for each spatial frequency (i. e., wave number) q.

The MSD of the ultraslow diffusion particle $x(t)$ can be derived from equation (7.17) as

$$\langle x^2(t)\rangle = 2D_\alpha E_\alpha^{-1}(t). \tag{7.19}$$

When $\alpha = 1$, equation (7.19) degenerates into the classical logarithmic diffusion law $\langle x^2(t)\rangle = 2D_1 \ln(t)$. It is noted that in the ultraslow diffusion equation (7.19), the MSD is infinite when $t = 0$, and its formula does not contain the classical Sinai anomalous diffusion law and its generalization. To circumvent these issues, instead we propose $k(t) = (E_\alpha^{-1}(1+t))^b$ as the structural function in the structural derivative definition equation (7.15); then equation (7.19) can be generalized into the following expression:

$$\langle x^2(t)\rangle = 2D_\alpha (E_\alpha^{-1}(1+t))^b, \tag{7.20}$$

where $b > 0$, and when $b > 4$, the defined diffusion process is a ballistic motion. Equation (7.20) is a general form of the logarithmic diffusion when $\alpha = 1$, and

$$\langle x^2(t)\rangle = 2D_1 (\ln(1+t))^b \tag{7.21}$$

is the Sinai diffusion when $b = 4$. Figure 7.2 presents the MSD for fixed diffusion coefficient $D_\alpha = 0.5$, $b = 1.0$ and four cases of α. It is observed from Figure 7.2 that the inverse ML diffusion proceeds more slowly than the logarithmic diffusion with increasing time.

The experimental data used in this study are taken from a previous study presented in [4]. We investigate the dynamics of two interacting particles in a disordered chain L. The two particles were put close to each other at the initial time. Then, the dynamics in the random system was characterized by short and long time scales. It has

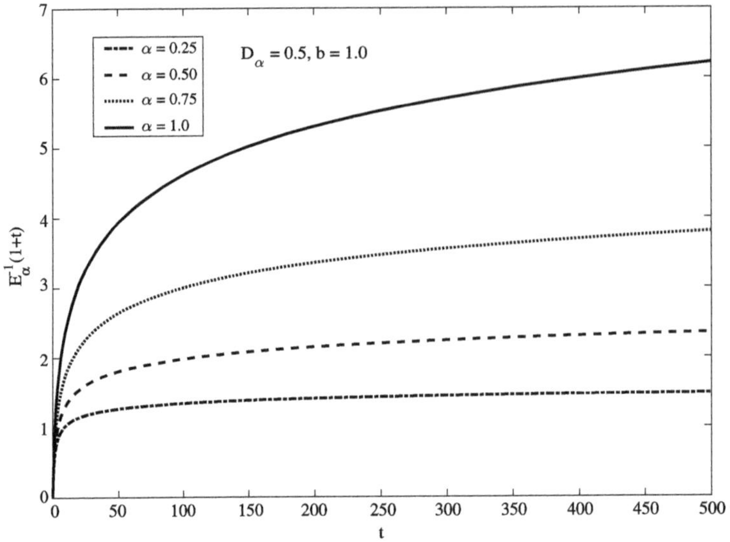

Figure 7.2: The inverse Mittag-Leffler diffusion described in equation (7.20) for fixed $D_\alpha = 0.5$, $b = 1.0$ and four cases of α.

been found that the interaction is a ballistic motion at the short time t_1 and manifests a very slow delocalization at the long time t_2 [4].

We choose two types of experimental data on the center of mass $R(t)$, which are assumed as the square root of the MSD from the diffusion point of view, i. e., $R(t) = (\langle x^2(t)\rangle)^{1/2}$. One case is at the short time $t_1 \leq 50$ in dimensionless unit with the size of chain $L = 200$ in dimensionless length unit and the localization length $L_1 = 200$. The other case is at the long time $t_2 \leq 10^5$ with the size of chain $L = 1024$ and the localization length $L_1 = 36$. Additional details of the data can be found in [4]. Both centers of mass $R(t)$ were respectively fitted by using the root square of the inverse ML diffusion law in equation (7.20) and the logarithmic diffusion law in equation (7.21). The parameters of the two laws were estimated by the least mean square method.

Table 7.1 provides the fitted parameters of the inverse ML diffusion [41]. The parameters D_α and b used in the logarithmic diffusion law in equation (7.21) are equal to the corresponding parameters of the ML diffusion law. Considering the results presented in Table 7.1, with fixed b, smaller α indicates a bigger gap between the inverse

Table 7.1: Parameters of the inverse Mittag-Leffler diffusion for the short and long time ultraslow diffusions.

Parameter	α	D_α	$b/2$
Short time	0.970	0.50	2.8
Long time	0.955	0.50	2.0

ML and the logarithmic diffusion laws. Instead, with fixed α, larger b shows a greater growing rate of $R(t)$ with increasing time. In addition, when $\alpha \neq 1$ the fitted $R(t)$ grows more slowly than in the logarithmic diffusion law. The values of the parameters also show that the diffusion process is the ballistic motion at the short time, and the Sinai ultraslow diffusion at the long time, which is consistent with the results from [4]. Overall, dynamics underlying ultraslow diffusion should exhibit smaller α with fixed b or smaller b with fixed α.

In Figures 7.3 and 7.4, we compare the experimental data with fits using the inverse ML and logarithmic diffusion laws, respectively, at the short and long time scales. From these plots, it is easily seen that both diffusion laws are correct when the time range is very small. With increasing time, the curves estimated by the logarithmic diffusion law, however, grow much faster than the data points, and they show a lower accuracy compared with the results estimated via the inverse ML diffusion law. From the perspective of accuracy, the inverse ML diffusion is a competitive alternative to describe the ultraslow diffusion.

It is worth noting that the underlying propagator of the inverse ML diffusion is the fundamental solution of the structural derivative diffusion equation based on the ML function. The equation generalizes the traditional diffusion equation and reflects the propagation of particles in regions with greater local interactions in the ultraslow diffusion than in the classical Brownian motion. This gives a physical interpretation of the inverse ML diffusion. The structural derivative appears a feasible mathematical tool to describe such ultraslow diffusion.

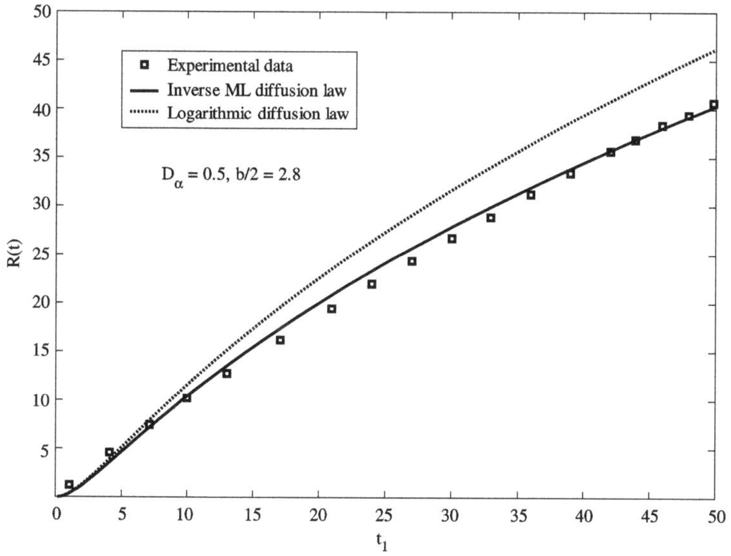

Figure 7.3: Plots of centers of mass $R(t)$ at the small time scale with fits using the root square of the inverse Mittag-Leffler and the logarithmic diffusion laws.

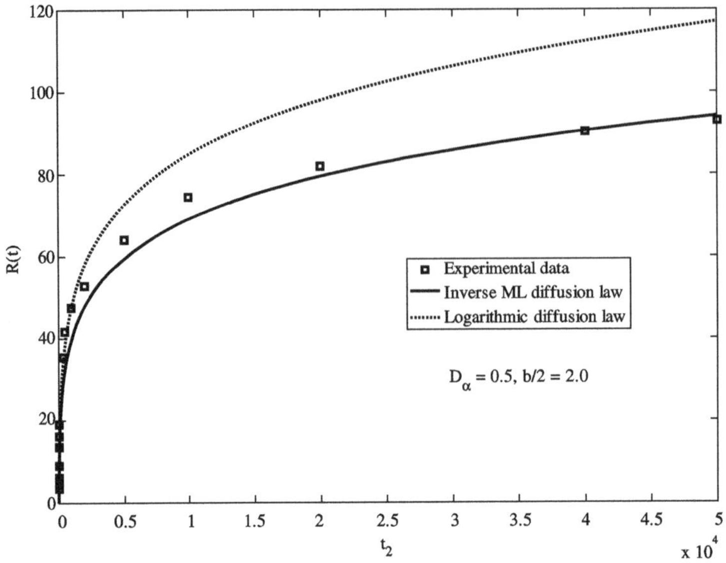

Figure 7.4: Plots of centers of mass $R(t)$ at the long time scale with fits using the root square of the inverse Mittag-Leffler and logarithmic diffusion laws.

In the above applications, the fitting results only demonstrate the usability of the inverse ML diffusion law, but the propagator has not been validated by experimental data. In future work, more experimental data containing the propagator information and the MSD might be considered to further test the proposed structural derivative model. It is also noted that in the present methodology, we select the inverse ML function as the structural function to model the ultraslow diffusion. It is possible that the other structural functions of clear physical significance, such as the stretched exponential function [27], the Bessel function [19], and their inverses, will be employed to construct local and non-local structural derivative models characterizing memory and non-locality of complex systems in either time or space. In addition, the definitions of structural derivative in space also need to be considered in real applications.

7.3.2 A spatial structural derivative model for ultraslow diffusion

In accordance with the local structural derivative, we establish the following spatial structural derivative model for ultraslow diffusion [194]:

$$\frac{\mathrm{d}p}{\mathrm{d}t} = K\frac{\mathrm{d}}{\mathrm{d}_s x}\left(\frac{\mathrm{d}p}{\mathrm{d}_s x}\right), \tag{7.22}$$

where K is the diffusion coefficient. When the structural function $f(x) = x$, equation (7.22) yields a Gaussian distribution, i. e.,

$$p(x,t) = \frac{1}{\sqrt{4\pi Kt}}\exp\left(-\frac{x^2}{4Kt}\right). \tag{7.23}$$

When $f(x) = x^{\beta}$, the solution of equation (7.22) is a stretched Gaussian distribution, i. e.,

$$p(x,t) = \frac{1}{\sqrt{4\pi Kt}}\exp\left(-\frac{x^{2\beta}}{4Kt}\right). \tag{7.24}$$

When $f(x) = e^x$, the corresponding structural derivative is

$$\frac{\mathrm{d}p}{\mathrm{d}_s x} = \lim_{x_1\to x}\frac{p(x_1,t)-p(x,t)}{e^{x_1}-e^{x}}, \tag{7.25}$$

and the corresponding solution of equation (7.22) can be derived as

$$p(x,t) = \frac{1}{\sqrt{4\pi Kt}}\exp\left(-\frac{e^{2x}}{4Kt}\right). \tag{7.26}$$

Substituting equation (7.26) into equation (7.22), we obtain

$$\frac{\mathrm{d}}{\mathrm{d}e^x}\left(\frac{\mathrm{d}p(x,t)}{\mathrm{d}e^x}\right) = -p(x,t)\left(\frac{1}{2Kt}-\frac{e^{2x}}{4K^2t^2}\right) = \frac{1}{K}\cdot\frac{\mathrm{d}}{\mathrm{d}t}. \tag{7.27}$$

Equation (7.26) is the solution of equation (7.22), in which the structural function is an exponential function. Equation (7.26) is a new kind of distribution, called the biexponential distribution in this work. The relationship between the structural function and the solution of the structural derivative diffusion equation in space is derived as

$$p(x,t) = \frac{1}{\sqrt{4\pi Kt}}\cdot\exp\left(-\frac{(f(x))^2}{4Kt}\right). \tag{7.28}$$

Generally speaking, the spatial structural derivative is a modeling strategy that can be employed in modeling ultraslow diffusion phenomena in complex fluids. The solution of the corresponding structural derivative diffusion equation constructed by the local structural derivative in space is statistical distribution, i. e., the probability density function.

Figure 7.5 is the probability density function described by Gaussian and biexponential distribution with $x > 0$, $t = 1$, and $K = 0.5$. From the simulation results, we can see that the biexponential distribution decreases more rapidly than Gaussian distribution in a short time. That means that compared with the probability of specific random variables falling in a specific range, the tailing phenomenon of the biexponential distribution is more evident. We numerically compute the MSD of the proposed ultraslow

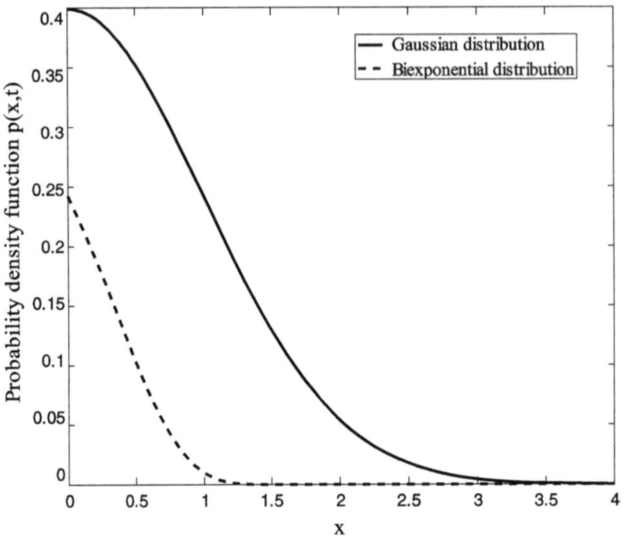

Figure 7.5: The probability density function with $t = 1$, $K = 0.5$.

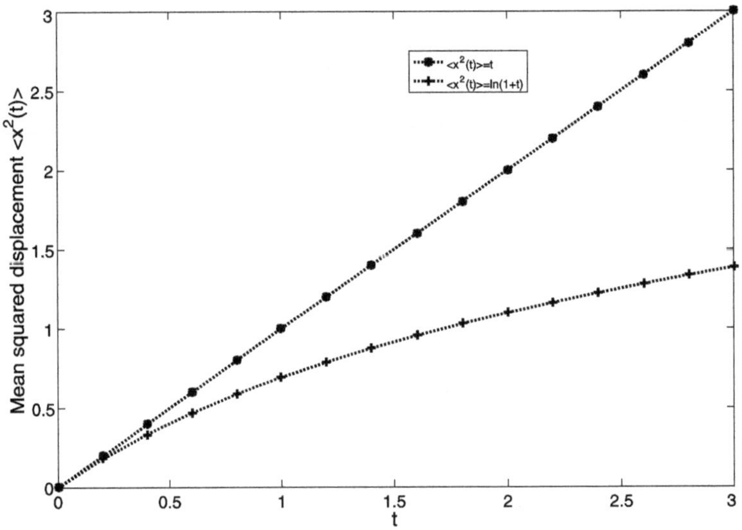

Figure 7.6: Schematic diagram of normal diffusion, sub-diffusion, super-diffusion, and the exponential structural derivative ultraslow diffusion, in which the proposed ultraslow diffusion and sub-diffusion are separated by the logarithmic ultraslow diffusion, $\langle x^2(t)\rangle = \ln(1+t)$, dotted with +, and the normal diffusion $\langle x^2(t)\rangle = t$ curve divides sub-diffusion and super-diffusion, dotted with *.

diffusion model, and we then explore the transient diffusion behavior by comparing with the normal diffusion, sub-diffusion, super-diffusion, and the proposed ultraslow diffusions. Figure 7.6 shows the differences of various diffusion processes.

In Figure 7.6, the yellow area represents the super-diffusion process. The corresponding MSD is $\langle x^2(t)\rangle = (t+1)^{\beta}$, $\beta > 1$. The blue and the green areas belong to the ultraslow diffusion and sub-diffusion, respectively.

The MSD of the proposed exponential function ultraslow diffusion can be derived from equation (7.28) as

$$\langle x^2(t)\rangle = \int_{-\infty}^{\infty} x^2 p(x,t)\mathrm{d}x = \frac{1}{\sqrt{4Kt\pi}}\int_{-\infty}^{\infty} x^2 \cdot \exp\left(-\frac{e^{2x}}{4Kt}\right)\mathrm{d}x. \tag{7.29}$$

Its analytical solution cannot be directly obtained; instead we define the MSD in $(0,+\infty)$ and calculate the following integral form:

$$\langle x^2(t)\rangle = \int_{0}^{\infty} x^2 p(x,t)\mathrm{d}x = \frac{1}{\sqrt{4Kt\pi}}\int_{0}^{\infty} x^2 \cdot \exp\left(-\frac{e^{2x}}{4Kt}\right)\mathrm{d}x. \tag{7.30}$$

Figure 7.7 shows the MSD of normal diffusion, logarithmic ultraslow diffusion, and the present exponential structural function ultraslow diffusion. We can observe from Figure 7.7 that the MSD of the proposed ultraslow diffusion increases slowlier with time than that of the logarithmic diffusion. Thus the local structural derivative diffusion equation with the structural function $f(x) = e^x$ in space is a mathematical modeling method to characterize a kind of ultraslow diffusion.

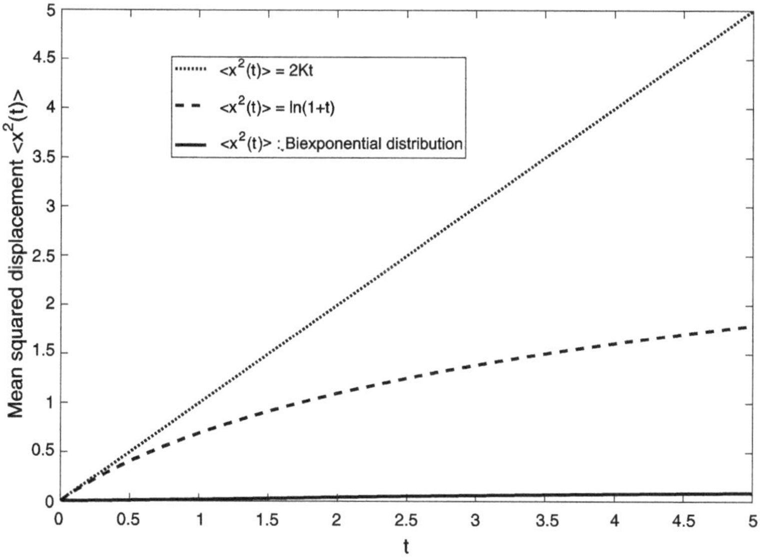

Figure 7.7: Mean squared displacement of normal diffusion, logarithmic ultraslow diffusion, and exponential function ultraslow diffusion with $K = 0.5$.

7.4 Structural derivative model for ultraslow creep

Ultraslow dynamics has attracted great interest in natural science and engineering [125, 143, 152]. It exists in the long-term effect of mechanical behaviors. However, little progress has been made in recent decades on the modeling of ubiquitous ultraslow creep phenomena, which have been observed in slow earthquakes [68, 84], glass systems [132], polymers [130], and civil engineering [71], just to mention a few. All these systems exhibit identical creep behavior, whose duration lasts for a long period.

The standard methods for modeling creep behavior are the Maxwell, the Kelvin, the linear solid, and the Burgers models and the finite chain of Kelvin elements or truncated Dirichlet series [11], which have been successfully applied to creep of polymers [164], soft mud [123], and concrete [107]. But the traditional constitutive models using the abovementioned mechanical elements are insufficient to effectively characterize the ultraslow creep due to their limited duration of scale. Thus, an effective model is highly desirable to describe the behavior over a large time range. In this section, the ultraslow creep model is constructed via the local structural derivative [40, 41]. The inverse ML function is employed as the structural function. The data on the durability and safety of Ultra-High Performance Concrete (UHPC), widely used in real-world engineering, are examined to verify the proposed model [76, 199].

Based on the structural derivative concept, we construct the local ultraslow creep model, in which the constitutive relation can be stated as

$$\sigma(t) = \eta \frac{d\varepsilon(t)}{d_s t}, \tag{7.31}$$

where $\sigma(t)$, $\varepsilon(t)$, and η denote the stress, the strain, and the consistency factor, respectively. The strain rate is defined as

$$\frac{d\varepsilon(t)}{d_s t} = \lim_{t_1 \to t} \frac{\varepsilon(t_1) - \varepsilon(t)}{E_\alpha^{-1}(t_1) - E_\alpha^{-1}(t)}, \tag{7.32}$$

where $E_\alpha^{-1}(t)$ is the inverse ML function with single parameter α. The ML function is defined with a power series $E_\alpha(t) = \sum_{k=0}^{\infty} \frac{t^k}{\Gamma(1+\alpha k)}$, in which Γ is the gamma function; $E_\alpha^{-1}(t)$ can be numerically calculated by the ML function. The inverse ML function degenerates into the classical logarithmic function when $\alpha = 1$.

According to the general definition of creep, the applied stress is constant, i. e., $\sigma(t) = \sigma_0$. From equations (7.31) and (7.32), the local structural derivative creep formula can be derived as [198]

$$\varepsilon(t) = \frac{\sigma_0}{\eta} \int_0^t \dot{E}_\alpha^{-1}(t) dt = \frac{\sigma_0}{\eta} E_\alpha^{-1}(t) + C. \tag{7.33}$$

Here, $\dot{E}_\alpha^{-1}(t)$ denotes the derivative of the inverse ML function. Based on the initial condition, $C = 0$. For simplicity $K = \frac{\sigma_0}{\eta}$. Then, equation (7.33) can be expressed as

$$\varepsilon(t) = K E_\alpha^{-1}(t). \tag{7.34}$$

When $t \to 0$, the strain is infinite. Also, different materials have different creep durations to reach equilibrium and the same sample also has different creep times under different conditions. Thus, a generalized structural function $k(t) = E_\alpha^{-1}(1+t)^\gamma$ tends to be favored. Ultimately, a more general ultraslow creep model is derived as

$$\varepsilon(t) = K(E_\alpha^{-1}(1+t))^\gamma. \tag{7.35}$$

Here, K, a constant, relates to the material parameters and test conditions; γ, the scaling exponent, is affected by the values of applied stress and its duration. It is noted that K has little impact on the magnitude of strain for a long-term range. Here K is assumed as finite.

The creep behavior of the UHPC is tested to verify the feasibility of the ultraslow creep model in equation (7.35). Data obtained from the literature and the fitting results are shown in Figure 7.8 and Figure 7.9. From the two figures, we see the simulated data of UHPC are in good agreement with the experimental data. The estimated parameters in the local structural derivative creep model are presented in Table 7.2. We can find from Table 7.2 that the magnitude of α increases with duration during creep. Meanwhile, a significant drop can be observed in the values of the scaling exponent γ (43 %), which shows that the extent of particle sliding and dislocation is sensitive to its duration in short time.

As a classical creep model, the Kelvin model is widely used to characterize the creep behaviors of concrete because of its simpler form and the presence of fewer pa-

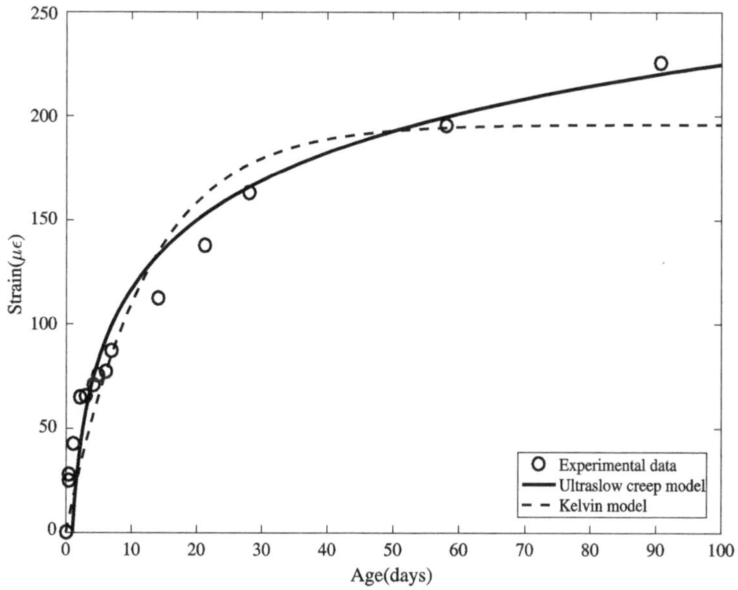

Figure 7.8: Comparison of results between experiments and simulations of the UHPC for a period of 90 days.

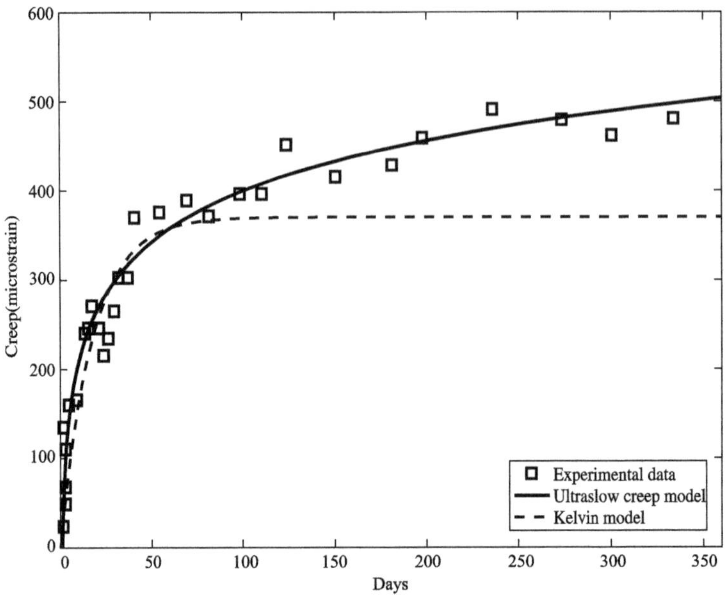

Figure 7.9: Comparison of results between experiments and simulations of the UHPC for a period of 340 days.

Table 7.2: Parameters of structural derivative creep model for the short-term and long-term scopes.

Parameter	α	γ
Short time	0.940	0.0145
Long time	0.948	0.0083

rameters [161]. In Figures 7.8 and 7.9, the fitting curves of the Kelvin model and the ultraslow model both at the short-term and long-term scopes are illustrated. Generally, the results can be divided into three stages. In the initial stage, the ultraslow creep model shows a higher creep rate than the Kelvin model. But in the transition stage, the Kelvin model grows much faster than the ultraslow creep model. Both of them show a good fitting result with experimental data in the initial and transition stages. In the last stage, the curve estimated by the Kelvin model inclines to a straight line and its corresponding results have lower accuracy compared with the experimental data. In contrast, the ultraslow creep model shows a distinct advantage for its ultraslow grow feature at the long-term scale. Thus, the ultraslow creep model can be considered as a better alternative method to depict long-term creep.

As a kind of typical porous material, the content of capillary pores plays an important role in determining the mechanical and durable properties of cement [12]. For the short-term creep, the higher content of air cavities gives rise to the larger deformation of microstructures when external stress is applied. Nevertheless, the content of pores reduces slightly when exposed to long-term stress, which contributes to the more compact and homogeneous micro-structure of UHPC. From the macroscopic view, the dense microstructures can supply higher resistance against external forces. Thus, the deformation of UHPC in the long term is smaller than that in the short term in terms of unit interval. The content of capillary pores in two cases can also be reflected by the value of γ, i. e., higher values of γ respond to higher content of capillary pores in the creep process.

Eventually, it should be noted that our study is mainly focused on the characterization of ultraslow creep. A qualitative analysis demonstrates the feasibility of the local structural derivative based on the inverse ML function. Although our methodology is supported by the creep data of UHPC, the internal relations between γ and material parameters (such as porosity and water/cement ratio) remain unknown. In the future, it is advised that more experimental data are acquired to investigate these quantitative relations. Furthermore, the structural derivative is promising and should be explored with more complex rheology phenomena based on different structural functions.

In this section, we present a local structural derivative method to investigate the ultraslow creep of UHPC. The inverse ML function is adopted as the structural function to construct the ultraslow creep model. Both short-term and long-term experimental data of UHPC can be well fitted by the ultraslow creep model. Compared with the traditional Kelvin model, the proposed model shows a better suitability and higher accuracy in long-term creep, in which the parameter γ is demonstrated to be a good marker to reflect the pore content of the selected samples. Our study also provides a framework for future studies to model ultraslow creep in more complex systems.

8 Perspectives

8.1 Model interpretation

In the previous chapters, the geometrical mechanism and statistics of Haudorff calculus and its PDF models with some potential applications are introduced. The present framework of Hausdorff calculus is physically sound and mathematically consistent as regards perspectives ranging from real applications to statistics and macromechanics to mesoscopic quantum mechanics. It should be pointed out that the hypotheses of fractal invariance and equivalence are presented in a somewhat heuristic way in the definition of Hausdorff calculus, which needs to be solidified in the future research.

When used the Hausdorff derivative model for a specific problem in engineering, e. g., solute transport in porous media; one issue is the simulation accuracy for the concentration of the tracer diffusion, and another issue is the physical mechanism of the model and its parameters [114]. The proposed Hausdorff derivative models are feasible and exhibit high precision for the stretched relaxation and the stretched Gaussian cases. The model parameters are needed to connect characteristics of the structure of the porous media, such as porosity, lacunarity, and tortuosity, which affect the solute transport process. The quantitative relationships between the model parameters and the structural features of the porous media should be discussed in the further.

8.2 Model selection

It is noted that both the traditional fractional calculus and the Hausdorff calculus are mathematical modeling formalisms underlying the scale space-time transforms. For instance, the inverse of the fractional time transform is also the kernel function in the definition of the fractional time derivative. However, the fractional derivatives in space and time are non-local, whereas the Hausdorff derivative is local. Both derivatives can give the generalized interpretation of diverse physical concepts on fractal space-time.

On the other hand, in recent years, the Tsallis distribution has also been a popular approach in the description of anomalous phenomena. Like the present stretched Gaussian distribution, this distribution was also a solution of the linear varying-coefficient Fokker–Planck equation of transport–diffusion type in which the standard local integer-order derivatives are used [156]. The corresponding Tsallis non-extensive thermodynamics is claimed to be capable to describe the long-range interacting systems and memory processes. The links and differences between the fractional and the Hausdorff derivatives for fractal space-time modeling are currently a subject under active study. Moreover, the physical mechanism and the application scope of each model are further needed to be summarized and explored.

https://doi.org/10.1515/9783110608526-008

8.3 Generalized fractal metrics

For more complicated media, the fractal structure cannot be well described by the power law fractal metrics. Instead, its metrics can be described by the generalized fractal metrics (GFM) in a non-Euclidean space, i. e., structal metrics or structural metrics [37],

$$\begin{cases} \Delta\hat{t} = \Delta T(t), \\ \Delta\hat{x} = \Delta Q(x), \end{cases} \tag{8.1}$$

where $T(t)$ and $Q(x)$ are structural functions. Equation (8.1) becomes the fractal metric when both $T(t)$ and $Q(x)$ are power functions. It is worthy of noting that the structal is different from the so-called multifractal [122], which depicts the fractal metric varies with time and space variables but remains a power function metric. Unlike the structural derivative, the varying-order fractional and Hausdorff derivatives can be used to describe multifractal systems.

It is known that the fractal has important applications in statistics and signal processing, such as $1/f$ noise analysis [105]. It is expected that the GFM has rich implications in statistics and probability as well. For example, the local time GFM have successfully been used to describe ultraslow diffusion with the inverse Mittag-Leffler function as the time structural function. The inverse Mittag-Leffler function characterizes the ultraslow power law decay (memory) and can find applications in time series analysis. The solution of the diffusion equation on GFM can also lead to new statistics.

Nowadays many artificial materials are invented and manufactured, such as mesomaterials, which have the metric to process specific-purpose functions. The structal concept and methodology could help to develop and analyze micro and mesostructures of such materials [137]. In addition, the complex network is another possible application field of the structal methodology, where the corresponding statistics methods may play an important role.

Bibliography

[1] Abbaszadeh M, Dehghan M, An improved meshless method for solving two-dimensional distributed order time-fractional diffusion-wave equation with error estimate, Numer Algorithms 75 (2017), 173–211.

[2] Abragam A, The Principles of Nuclear Magnetism, Oxford University Press, Oxford, 1961.

[3] Alves L G, Scariot D B, Guimarães R R, et al., Transient super-diffusion and long-range correlations in the motility patterns of trypanosomatid flagellate protozoa, PLoS ONE 11 (2016), e0152092.

[4] Arias S D T, Waintal X, Pichard J, Two interacting particles in a disordered chain III: dynamical aspects of the interplay disorder-interaction, Eur Phys J B 10 (1999), 149–158.

[5] Arimitsu T, Arimitsu N, Multifractal analysis of fluid particle accelerations in turbulence, Physica D 193 (2004), 218–230.

[6] Balankin A S, A continuum framework for mechanics of fractal materials I: from fractional space to continuum with fractal metric, Eur Phys J B 88 (2015), 1–13.

[7] Balankin A S, Bory-Reyes J, Shapiro M, Towards a physics on fractals: differential vector calculus in three-dimensional continuum with fractal metric, Physica A 444 (2016), 345–359.

[8] Balankin A S, Elizarraraz B E, Map of fluid flow in fractal porous medium into fractal continuum flow, Phys Rev E 85 (2012), 056314.

[9] Banks D S, Fradin C, Anomalous diffusion of proteins due to molecular crowding, Biophys J 89 (2005), 2960–2971.

[10] Barnes H A, Thixotropy-a review, J Non-Newton Fluid 70 (1997), 1–33.

[11] Barnes H A, Hutton J F, Walters K, An Introduction to Rheology, Elsevier, Amsterdam, 1993.

[12] Basheer L, Kropp J, Cleland D J, Assessment of the durability of concrete from its permeation properties: a review, Constr Build Mater 15 (2001), 93–103.

[13] Beck C, Non-extensive statistical mechanics approach to fully developed hydrodynamic turbulence, Chaos Solitons Fractals 13 (2002), 499–506.

[14] Beck C, Superstatistics in hydrodynamic turbulence, Physica D, 193 (2004), 195–207.

[15] Beck C, Stretched exponentials from superstatistics, Physica A 365 (2005), 96–101.

[16] Bennett K M, Schmainda K M, Bennett R T, et al., Characterization of continuously distributed cortical water diffusion rates with a stretched exponential model, Magn Reson Med 50 (2003), 727–734.

[17] Blum R S, Kozick R J, Sadler B M, An adaptive spatial diversity receiver for non-Gaussian interference and noise, IEEE Trans Signal Process 47 (1999), 2100–2111.

[18] Bouchaud J P, Georges A, Anomalous diffusion in disordered media: statistical mechanisms, models and physical applications, Phys Rep 195 (1990), 127–293.

[19] Bowman F, Introduction to Bessel functions, Nature 143 (1939), 356.

[20] Brauns E B, Madaras M L, Coleman R S, et al., Complex local dynamics in DNA on the picosecond and nanosecond time scales, Phys Rev Lett 88 (2002), 158101.

[21] Cai W, Chen W, Application of scaling transformation to characterizing complex rheological behaviors and fractal derivative modeling, Rheol Acta 57 (2018), 43–50.

[22] Cai W, Chen W, Wang F, Three-dimensional Hausdorff derivative diffusion model for isotropic/anisotropic fractal porous media, Therm Sci 22 (2018), 1–6.

[23] Cai W, Chen W, Xu W, Characterizing the creep of viscoelastic materials by fractal derivative models, Int J Non-Linear Mech 87 (2016), 58–63.

[24] Cai W, Chen W, Xu W, The fractal derivative wave equation: application to clinical amplitude/velocity reconstruction imaging, J Acoust Soc Am 143 (2018), 1559–1566.

https://doi.org/10.1515/9783110608526-009

[25] Callaghan P T, Principles of Nuclear Magnetic Resonance Microscopy, Oxford University Press, Oxford, 1991.
[26] Caputo M, Fabrizio M, A new definition of fractional derivative without singular kernel, Prog Fract Differ Appl 1 (2015), 73–85.
[27] Cardona M, Chamberlin R V, Marx W, The history of the stretched exponential function, Ann Phys-Berlin 16 (2007), 842–845.
[28] Chambers J M, Mallows C, Stuck B, A method for simulating stable random variables, J Am Stat Assoc 71 (1976), 340–344.
[29] Chapman C H, Orcutt J A, Least-square fitting of marine seismic refraction data, Geophys J Int 82 (2007), 339–374.
[30] Chen S, Mulgrew B, Hanzo L, Least bit error rate adaptive nonlinear equalizers for binary signaling, IEEE Pers Commun 150 (2003), 29–36.
[31] Chen W, Time-space fabric underlying anomalous diffusion. Chaos Solitons Fractals 28 (2006), 923–929.
[32] Chen W, A speculative study of 2/3-order fractional Laplacian modeling of turbulence: some thoughts and conjectures, Chaos 16 (2006), 023126.
[33] Chen W, Singular boundary method: a novel, simple, meshfree, boundary collocation numerical method, Chin J Solid Mech 30 (2009), 592–599.
[34] Chen W, Implicit calculus modeling for simulation of complex scientific and engineering problems, Comput Aided E 23 (2014), 1–6.
[35] Chen W, Fractal geometric theory for Hausdorff calculus and fractional calculus models. Inter Comput-Aided E 27 (2017), 1–6.
[36] Chen W, Fractal analysis of Hausdorff calculus and fractional calculus models, Comput-Aided 26 (2017), 1–5.
[37] Chen W, Non-power-function metric: a generalized fractal, Math Phys 2017, viXra:1612.0409.
[38] Chen W, Fu Z J, Chen C S, Recent Advances in Radial Basis Function Collocation Methods, Springer, 2014.
[39] Chen W, Hei X, Liang Y, A fractional structural derivative model for ultraslow diffusion, Appl Math Mech 37 (2016), 599–608.
[40] Chen W, Liang Y, Hei X, Local structural derivative and its applications, Acta Mech Solida Sin 37 (2016), 456–460.
[41] Chen W, Liang Y, Hei X, Structural derivative based on inverse Mittag-Leffler function for modeling ultraslow diffusion, Fract Calc Appl Anal 19 (2016), 1250–1261.
[42] Chen W, Liang Y, Hu S, et al., Fractional derivative anomalous diffusion equation modeling prime number distribution, Fract Calc Appl Anal 18 (2015), 789–798.
[43] Chen W, Liang Y, Yang X, et al., Soil infiltration rates and hydrology model classifications based on the Hausdorff fractal derivative Richards equation, Appl Math Mech Eng 39 (2018), 77–82.
[44] Chen W, Pang G, A new definition of fractional Laplacian with application to modeling three-dimensional nonlocal heat conduction, J Comput Phys 309 (2016), 350–367.
[45] Chen W, Sun H, Zhang X, Korošak D, Anomalous diffusion modeling by fractal and fractional derivatives, Comput Math Appl 59 (2010), 1754–1758.
[46] Chen W, Sun H G, Multiscale statistical model of fully-developed turbulence particle accelerations, Mod Phys Lett B 23 (2009), 449–452.
[47] Chen W, Wang F, Zheng B, et al., Non-Euclidean distance fundamental solution of Hausdorff derivative partial differential equations, Eng Anal Bound Elem 84 (2017), 213–219.
[48] Chen W, Wang F Z, A method of fundamental solutions without fictitious boundary, Eng Anal Bound Elem 34 (2010), 530–532.
[49] Chen X, Concise History of Statistics, Hunan Education Publishing House, Changsha, 2002.

[50] Cheng A D, Multiquadric and its shape parameter-a numerical investigation of error estimate, condition number, and round-off error by arbitrary precision computation, Eng Anal Bound Elem 36 (2012), 220–239.

[51] Cheng H D, Cheng D T, Heritage and early history of the boundary element method, Eng Anal Bound Elem 29 (2005), 268–302.

[52] Chevalier Y, Vinh J T, Mechanics of Viscoelastic Materials and Wave Dispersion, 2013.

[53] Chrysostomos L, Shao M, Signal processing with alpha-stable distributions and applications, Comput Stat Data Anal 22 (1996), 333–334.

[54] Cristescu N, A general constitutive equation for transient and stationary creep of rock salt, Int J Rock Mech Min 30 (1993), 125–140.

[55] Deng W, Li C, Lv J, Stability analysis of linear fractional differential system with multiple time delays, Nonlinear Dyn 48 (2007), 409–416.

[56] Dobierzewska-Mozrzymas E, Szymczak G, Biegański P, Pieciul E, Lévy's distributions for statistical description of fractal structures; discontinuous metal films on dielectric substrates, Physica B 337 (2003), 79–86.

[57] Eftekhari M, Fatemi A, Creep behavior and modeling of neat, talc-filled, and short glass fiber reinforced thermoplastics, Composites, Part B 97 (2016), 68–83.

[58] Ehsani A, Mahjani M G, Bordbar M, et al., Electrochemical study of anomalous diffusion and fractal dimension in poly ortho aminophenol electroactive film: comparative study, J Electroanal Chem 710 (2013), 29–35.

[59] Fan W P, Jiang X Y, Parameters estimation for a one-dimensional time fractional thermal wave equation with fractional heat flux conditions, Acta Phys Sin 63 (2014), 140202.

[60] Fang X J, Xu Y C, Shi T Y, Zhang L, Li Z F, Stress and continuous relaxation spectrum porcine cornea after LASIK. Int J Ophthalmol 7 (2007), 626–629.

[61] Fasshauer G E, Newton iteration with multiquadrics for the solution of nonlinear PDEs, Comput Math Appl 43 (2002), 423–438.

[62] Fasshauer G E, Meshfree Approximation Methods with Matlab: (with CD-ROM), World Scientific Publishing Co Inc. 2007.

[63] Fasshauer G E, Zhang J G, On choosing "optimal" shape parameters for RBF approximation, Numer Algorithms 45 (2007), 345–368.

[64] Flandrin P, Wavelet analysis and synthesis of fractional Brownian motion, IEEE Trans Inf Theory 38 (1992), 910–917.

[65] Franke R, Scattered data interpolation: tests of some methods, Math Comput 38 (1982), 181–200.

[66] Freund J B, Ewoldt R H, Quantitative rheological model selection: good fits versus credible models using Bayesian inference, J Rheol 59 (2015), 667–701.

[67] Fuentes M A, Toral R, Wio H S, Enhancement of stochastic resonance: the role of non Gaussian noises, Physica A 295 (2001), 114–122.

[68] Furuya M, Satyabala S P, Slow earthquake in Afghanistan detected by InSAR, Geophys Res Lett 35 (2008), 160–162.

[69] Galucio A C, Deu J F, Ohayon R, A fractional derivative viscoelastic model for hybrid active-passive damping treatments in time domain-application to sandwich beams, J Intell Mater Syst Struct 16 (2005), 33–45.

[70] Gao Q, Srinivasan G, Magin R L, Zhou X J, Anomalous diffusion measured by a twice-refocused spin echo pulse sequence: analysis using fractional order calculus, J Magn Reson Imaging 33 (2011), 1177–1183.

[71] Garas V Y, Kurtis K E, Kahn L F, Creep of UHPC in tension and compression: effect of thermal treatment, Cem Concr Compos 34 (2012), 493–502.

[72] R. Garrappa, Predictor-corrector pece method for fractional differential equations, https://cn.mathworks.com/matlabcentral/profile/authors/2361481-roberto-garrappa.

[73] Gatto R G, Chu Y, Ye A Q, et al., Analysis of YFP(J16)-R6/2 reporter mice and post-mortem brains reveals early pathology and increased vulnerability of callosal axons in Huntington's disease, Hum Mol Genet 24 (2015), 5285–5298.

[74] Godinho L, Branco F, Formulation of Kansa's method in the frequency domain for the analysis of transient heat conduction, Int J Numer Methods Heat Fluid Flow 24 (2014), 1437–1453.

[75] Gorenflo R, Mainardi F, Fractional calculus: integral and differential equations of fractional order, Mathematics 49 (2008), 277–290.

[76] Graybeal B, Material property characterization of ultra-high performance concrete, Creep 1 (2006), 887–894.

[77] Guimarães F G, Saldanha R R, Mesquita R C, et al., A meshless method for electromagnetic field computation based on the multiquadric technique, IEEE Trans Magn 43 (2007), 1281–1284.

[78] Han C, Zhao L, Zhong S, et al., A comparison of high b-value vs standard b-value diffusion-weighted magnetic resonance imaging at 3.0 T for medulloblastomas, Br J Radiol 88 (2015), 0150220.

[79] G. H. Hardy, Gösta Mittag-Leffler, J Lond Math Soc 1 (1928), 156–160.

[80] Hardy R L, Multiquadric equations of topography and other irregular surfaces, J Geophys Res 76 (1971), 1905–1915.

[81] Hartnett J, Cho Y, Hartnett J, Non-Newtonian Fluids, Handbook of Heat Transfer, 3rd ed, New York, 1998.

[82] Hasan O A, Boyce M C, A constitutive model for the nonlinear viscoelastic viscoplastic behavior of glassy polymers, Polym Eng Sci 35 (1995), 331–344.

[83] He L F, Cui Y Y, Zhang T Q, Zhang G, Song Y, Analysis of weak signal detection based on tri-stable system under Levy noise, Chin Phys B 25 (2016), 85–94.

[84] Heki K, Miyazaki S I, Tsuji H, Silent fault slip following an interplate thrust earthquake at the Japan Trench, Nature 386 (1997), 595–598.

[85] Herrera I, Pinder G F, Mathematical Modeling in Science and Engineering: An Axiomatic Approach, Wiley & Sons, 2012.

[86] Hilfer R, Seybold H J, Computation of the generalized Mittag-Leffler function and its inverse in the complex plane, Integral Transforms Spec Funct 17 (2006), 637–652.

[87] Hooshyar M, Wang D, An analytical solution of Richards' equation providing the physical basis of SCS curve number method and its proportionality relationship, Water Resour Res 52 (2016), 6611–6620.

[88] Hu F, Zhu W Q, Chen L C, Stochastic fractional optimal control of quasi-integrable Hamiltonian system with fractional derivative damping, Nonlinear Dyn 70 (2012), 1459–1472.

[89] Huang X, Study on ANN noise adaptability in application of industry process characteristics mining, Adv Mater Res. 462 (2012), 635–640.

[90] Hummel A, Wesche K, Brand W, Rüsch H, Kordina K, Hilsdorf H, Versuche über das Kriechen unbewehrten Betons: Der Einfluss der Zementart, des Wasser-Zement-Verhältnisses und des Belastungsalters auf das Kriechen von Beton. Wilhelm Ernst & Sohn, 1962.

[91] Ingo C, Magin R L, Colon-Perez L, Triplett W, Mareci T H, On random walks and entropy in diffusion-weighted magnetic resonance imaging studies of neural tissue, Magn Reson Med 71 (2014), 617–627.

[92] Ingo C, Magin R L, Parrish T B, New insights into the fractional order diffusion equation using entropy and kurtosis, Entropy 16 (2014), 5838–5852.

[93] Ingo C, Sui Y, Chen Y, Ronen I, Parsimonious continuous time random walk models and kurtosis for diffusion in magnetic resonance of biological tissue, Front Phys 3 (2015) 11.

[94] Jensen J H, Helpern J A, Ramani A, Lu H Z, Kaczynski K, Diffusional kurtosis imaging: the quantification of non-Gaussian water diffusion by means of magnetic resonance imaging, Magn Reson Med 53 (2005), 1432–1440.

[95] Johansson R, Direct adaptive control-global Lyapunov stability and exponential convergence, Int J Control 50 (2007), 859–869.

[96] Jost W, Diffusion in Solids, 3rd ed, Academic Press, New York, 1960.

[97] Jumarie G, On the representation of fractional Brownian motion as an integral with respect to $(dt)^a$, Appl Math Lett 18 (2005), 739–748.

[98] Kanno R, Representation of random walk in fractal space-time, Physica A 248 (1998), 165–175.

[99] Kansa E J, Multiquadrics-A scattered data approximation scheme with applications to computational fluid-dynamics-II solutions to parabolic, hyperbolic and elliptic partial differential equations, Comput Math Appl 19 (1990), 147–161.

[100] Kansa E J, Multiquadrics-A scattered data approximation scheme with applications to computational fluid-dynamics-I surface approximations and partial derivative estimates, Comput Math Appl 19 (1990), 127–145.

[101] Karageorghis A, Chen C S, Liu X Y, Kansa-RBF algorithms for elliptic problems in axisymmetric domains, SIAM J Sci Comput 38 (2016), 435–470.

[102] Khalil H K, Nonlinear Systems, 3rd ed, PrenticeHall, Inc., Upper Saddle River, NJ, 2002.

[103] Kindelan M, Bernal F, González-Rodríguez P, et al., Application of the RBF meshless method to the solution of the radiative transport equation, J Comput Phys 229 (2010), 1897–1908.

[104] Kolchigin N N, Pivnenko S N, Numerical modeling of measurements of dielectric material characteristics using non-sinusoidal signals, Proc Int Conf Digital Audio Effects 1 (1998), 346–348.

[105] Kolek A, Multifractality and 1/f noise in the two-component random resistor network, Phys Rev B, Condens Matter 45 (1992), 205–208.

[106] Lancaster P, Šalkauskas K, Curve and Surface Fitting: An Introduction, Academic Press, 1986.

[107] Lee A R, Markwick A, The mechanical properties of bituminous surfacing materials under constant stress, J Soc Chem Ind 56 (1937), 146–156.

[108] Li C, Peng G, Chaos in Chen's system with a fractional order, Chaos Solitons Fractals 22 (2004), 443–450.

[109] Li C, Shi Q, Li B, Experimental investigations and derivation of model for the relaxation of saturated soft clays, Soil Eng Found 15 (1989), 38–40.

[110] Li J, Ostoja-Starzewski M, Fractal solids, product measures and fractional wave equations, Proc R Soc A, Math Phys 465 (2009), 2521–2536.

[111] Li J, Ostoja-Starzewski M, Comment on "Hydrodynamics of fractal continuum flow" and "Map of fluid flow in fractal porous medium into fractal continuum flow", Phys Rev E 88 (2013), 057001.

[112] Li Y, Chen Y Q, Podlubny I, Mittag-Leffler stability of fractional order nonlinear dynamic systems, Automatica 45 (2009), 1965–1969.

[113] Li Y, Chen Y Q, Podlubny I, Stability of fractional-order nonlinear dynamic systems: Lyapunov direct method and generalized Mittag-Leffler stability, Comput Math Appl 59 (2010), 1810–1821.

[114] Liang Y, Su N, Chen W, Yang X, A time-space Hausdorff fractal model for non-Fickian transport in porous media, arXiv:1801.09352.

[115] Liang Y, Ye A Q, Chen W, et al., A fractal derivative model for the characterization of anomalous diffusion in magnetic resonance imaging, Commun Nonlinear Sci 39 (2016), 529–537.

[116] Liang Y J, Chen W, A survey on computing Lévy stable distributions and a new MATLAB toolbox, Signal Process 93 (2013), 242–251.

[117] Liang Y J, Chen W, Bridge fatigue life prediction using Mittag-Leffler distribution, Fatigue Fract Eng Mater Struct 37 (2014), 255–264.
[118] Liao W H, Roebel A, Su W Y, On stretching Gaussian Noises with the Phase Vocoder, Proc Int Conf Digital Audio Effects 9 (2012), 17–21.
[119] Lin G, An effective phase shift diffusion equation method for analysis of PFG normal and fractional diffusions, J Magn Reson 259 (2015), 232–240.
[120] Lomholt M A, Lizana L, Metzler R, et al., Microscopic origin of the logarithmic time evolution of aging processes in complex systems, Phys Rev Lett 110 (2013), 208301.
[121] Long H, Ma C, Shimizu Y, Least squares estimators for stochastic differential equations driven by small Lévy noises, Stoch Process Appl 8 (2016), 1–21.
[122] Lopes R, Betrouni N, Fractal and multifractal analysis: a review, Med Image Anal 13 (2009), 634–649.
[123] Maa J, Mehta A J, Soft mud properties: Voigt model, J Waterw Port C-ASCE. 114 (1988), 765–770.
[124] Machado J T, Kiryakova V, Mainardi F, Recent history of fractional calculus, Commun Nonlinear Sci 16 (2011), 1140–1153.
[125] Macovez R, Mitsari E, Zachariah M, et al., Ultraslow dynamics of water in organic molecular solids, J Phys Chem C 118 (2014), 4941–4950.
[126] Magin R L, Akpa B S, Neuberger T, Webb A G, Fractional order analysis of Sephadex gel structures: NMR measurements reflecting anomalous diffusion, Commun Nonlinear Sci 16 (2011), 4581–4587.
[127] Magin R L, Ingo C, Entropy and information in a fractional order model of anomalous diffusion, Syst Identif 45 (2012), 428–433.
[128] Magin R L, Ingo C, Colon-Perez L, Triplett W, Mareci T H, Characterization of anomalous diffusion in porous biological tissues using fractional order derivatives and entropy, Microporous Mesoporous Mater 178 (2013), 39–43.
[129] Mainardi F, Fractional Calculus and Waves in Linear Viscoelasticity: An Introduction to Mathematical Models, World Scientific, 2010.
[130] Mandare P, Winter H H, Ultraslow dynamics in asymmetric block copolymers with nanospherical domains, Colloid Polym Sci 284 (2006), 1203–1210.
[131] Manolis G D, Polyzos D D, Recent Advances in Boundary Element Methods, Springer Netherlands, 2009.
[132] Martinez-Garcia J C, Rzoska S J, Drozd-Rzoska A, Martinez-Garcia J, A universal description of ultraslow glass dynamics, Nat Commun 4 (2013), 1823.
[133] Matan K, Williams R B, Witten T A, et al., Crumpling a thin sheet, Phys Rev Lett 88 (2002), 076101.
[134] Metzler R, Jeon J H, Cherstvy A G, Barkai E, Anomalous diffusion models and their properties: non-stationarity, non-ergodicity, and ageing at the centenary of single particle tracking, Phys Chem Chem Phys 16 (2014), 24128.
[135] Metzler R, Klafter J, The random walk's guide to anomalous diffusion: a fractional dynamics approach, Phys Rep 339 (2000), 1–77.
[136] Metzler R, Nonnenmacher T F, Fractional relaxation processes and fractional rheological models for the description of a class of viscoelastic materials, Int J Plast 19 (2003), 941–959.
[137] Miao Y, Lu G, Liu X, et al., Mo-functionalized MCF meso-material and its catalytic performance for epoxidation of propylene by cumene hydroperoxide, Microporous Mesoporous Mater 122 (2009), 55–60.
[138] Middleton D, Non-Gaussian noise models in signal processing for telecommunications: new methods and results for class A and class B noise models, IEEE Trans Inf Theory 45 (1999), 1129–1149.

[139] Milovanov A V, Rypdal K, Rasmussen J J, Stretched exponential relaxation and ac universality in disordered dielectrics. Phys Rev B 76 (2007), 3398–3407.
[140] Mishra S K, Singh V P, Soil Conservation Service Curve Number (SCS-CN) Methodology, Springer Netherlands, 2003.
[141] Moës, N, Dolbow J, Belytschko T, A finite element method for crack growth without remeshing, Int J Numer Methods Eng 46 (2015), 131–150.
[142] Mordant N, Crawford A M, Bodenschatz E, Experimental Lagrangian acceleration probability density function measurement, Physica D 193 (2003), 245–251.
[143] Nalbach P, Thorwart M, Ultraslow quantum dynamics in a sub-Ohmic heat bath, Phys Rev B, Condens Matter 81 (2010), 054308.
[144] Nasri A, Schober R, Performance of BICM-SC and BICM-OFDM systems with diversity reception in non-Gaussian noise and interference, IEEE Trans Commun 57 (2009), 3316–3327.
[145] Nasri A, Schober R, Ma Y, Unified asymptotic analysis of linearly modulated signals in fading, non-Gaussian noise and interference, IEEE Trans Commun 56 (2008), 980–990.
[146] Norris D G, Niendorf T, Leibfritz D, Healthy and infarcted brain tissues studied at short diffusion times: the origins of apparent restriction and the reduction in apparent diffusion coefficient, NMR Biomed 7 (2010) 304–310.
[147] Oates K M, Krause W E, Jones R L, et al., Rheopexy of synovial fluid and protein aggregation, J R Soc Interface 3 (2006) 167–174.
[148] Paulsen J D, Nagel S R, A model for approximately stretched-exponential relaxation with continuously varying stretching exponents, J Stat Phys 167 (2017), 749–762.
[149] Paxinos G, Franklin K B J, The Mouse Brain in Stereotaxic Coordinates, Gulf Professional Publishing, 2004.
[150] Podlubny I, Geometric and physical interpretation of fractional integration and fractional differentiation, Fract Calc Appl Anal 5 (2001), 230–237.
[151] Porta A L, Voth G A, Crawford A M, et al., Fluid particle accelerations in fully developed turbulence, Nature 409 (2001), 1017–1019.
[152] Ramos L, Cipelletti L, Ultraslow dynamics and stress relaxation in the aging of a soft glassy system, Phys Rev Lett 87 (2001), 245503.
[153] Ren F Y, Wang J, Lv L J, Pan H, Qiu W Y, Effect of different waiting time processes with memory to anomalous diffusion dynamics in an external force fields, Physica A 417 (2015), 202–214.
[154] Ren F Y, Xu Y, Qiu W Y, Liang J R, Universality of stretched Gaussian asymptotic diffusion behavior on biased heterogeneous fractal structure in external force fields, Chaos Solitons Fractals 24 (2005), 273–278.
[155] Reyes-Marambio J, Moser F, Gana F, Severino B, Calderón-Muñoz W R, Palma-Behnke R, Estevez P A, Orchard M, Cortés M, A fractal time thermal model for predicting the surface temperature of air-cooled cylindrical Li-ion cells based on experimental measurements, J Power Sources 306 (2016), 636–645.
[156] Ribeiro M S, Nobre F D, Curado E M F, Classes of n-dimensional nonlinear Fokker–Planck equations associated to Tsallis entropy, Entropy 13 (2011), 1928–1944.
[157] Rippa S, An algorithm for selecting a good value for the parameter c in radial basis function interpolation, Adv Comput Math 11 (1999), 193–210.
[158] Sanders L, Lomholt M, Lizana L, et al., Severe slowing-down and universality of the dynamics in disordered interacting many-body systems: ageing and ultraslow diffusion, New J Phys 16 (2014), 113050.
[159] Schiessel H, Metzler R, Blumen A, Nonnenmacher T F, Generalized viscoelastic models: their fractional equations with solutions, J Phys A Math Theory 28 (1995) 6567–6584.
[160] Schumer R, Meerschaert M M, Baeumer B, Fractional advection dispersion equations for modeling transport at the Earth surface, J Geophys Res Earth Surf 114 (2009), F00A07.

[161] Schutter G D, Degree of hydration based Kelvin model for the basic creep of early age concrete, Mater Struct 32 (1999) 260–265.

[162] Scott Blair G W, An application of the theory of quasi-properties to the treatment of anomalous stress-strain relations, Philos May Ser 40 (1949), 80–94.

[163] Segev R, Benveniste M, Hulata E, et al., Long term behavior of lithographically prepared in vitro neuronal networks, Phys Rev Lett 88 (2002), 118102.

[164] Shaw M T, Macknight W J, Introduction to Polymer Viscoelasticity, 3rd ed, John Wiley & Sons, Inc., 1983.

[165] Shlesinger M F, West B J, Klafter J, Lévy dynamics of enhanced diffusion: application to turbulence, Phys Rev Lett 58 (1987), 1100–1103.

[166] Sikora M, Adamczyk G, Krystyjan M, et al., Thixotropic properties of normal potato starch depending on the degree of the granules pasting, Carbohydr Polym 121 (2015), 254–264.

[167] Sinai Y G, The limiting behavior of a one-dimensional random walk in a random medium, Theory Probab Appl 27 (1983), 256–268.

[168] Sokolov I M, Ebelling W, Dybiec B, Harmonic oscillator under Lévy noise: unexpected properties in the phase space, Phys Rev E 83 (2011), 041118.

[169] Solomon T H, Weeks E R, Swinney H L, Observation of anomalous diffusion and Lévy flights in a two-dimensional rotating flow, Phys Rev Lett 71 (1994), 3975–3978.

[170] Song R, Muliana A H, Palazotto A, An empirical approach to evaluate creep responses in polymers and polymeric composites and determination of design stresses, Compos Struct 148 (2016), 207–223.

[171] Sorrentino S, Fasana A, Finite element analysis of vibrating linear systems with fractional derivative viscoelastic models, J Sound Vib 299 (2007), 839–853.

[172] Stanley H E, Havlin S, Generalisation of the Sinai anomalous diffusion law, J Phys A, Math Theor 20 (1987), L615.

[173] Stepišnik J, Analysis of NMR self-diffusion measurements by a density matrix calculation, Physica B 104 (1982), 350–364.

[174] Su X, Chen W, Xu W, et al., Characterizing the rheological behaviors of non-Newtonian fluid via a viscoelastic component: fractal dashpot. Adv Mech Eng 9 (2017), 168781401769976.

[175] Su X, Xu W, Chen W, Numerical study for laminar flow of non-Newtonian fluid based on fractal derivative, Chin J Theor Appl Mech 49 (2017), 1020–1028.

[176] Sugita S, Yagi M, Itoh S, Itoh K, Bohm-like dependence of transport in scrape-off layer plasmas, J Phys Soc Jpn 81 (2012), 044501.

[177] Sun H, Meerschaert M M, Zhang Y, Zhu J, Chen W, A fractal Richards' equation to capture the non-Boltzmann scaling of water transport in unsaturated media, Adv Water Resour 52 (2013), 292–295.

[178] Tarasov V E, Anisotropic fractal media by vector calculus in non-integer dimensional space, J Math Phys 55 (2014), 1199–1208.

[179] Toivonen J, Merisaari H, Pesola M, et al., Mathematical models for diffusion-weighted imaging of prostate cancer using b values up to 2000 s/mm^2: correlation with Gleason score and repeatability of region of interest analysis, Magn Reson Med 74 (2015), 1116–1124.

[180] Tseng C C, Lee S L, Designs of two-dimensional linear phase FIR filters using fractional derivative constraints, Signal Process 93 (2013), 1141–1151.

[181] Tsihrintzis G A, Nikias C L, Fast estimation of the parameters of alpha-stable impulsive interference, IEEE Trans Signal Process 44 (1995), 1492–1503.

[182] Vaknin A, Ovadyahu Z, Pollak M, Aging effects in an Anderson insulator, Phys Rev Lett 84 (2000), 3402.

[183] Viertiö-Oja H, Maja V, Särkelä M, et al., Description of the entropy algorithm as applied in the Datex-Ohmeda *S*/5 entropy module, Acta Anaesth Scand 48 (2004), 154–161.

[184] Wang B C, Wei Y H, Sun Y H, Generate random number by using acceptance rejection method, J Chongqing Norm Univ (Natural science) 30 (2013), 86–91.
[185] Wang F, Chen W, Gu Y, Boundary element analysis of inverse heat conduction problems in 2D thin-walled structures, Int J Heat Mass Transf 91 (2015), 1001–1009.
[186] Wang F, Chen W, Qu W, Gu Y, A BEM formulation in conjunction with parametric equation approach for three-dimensional Cauchy problems of steady heat conduction, Eng Anal Bound Elem 63 (2016), 1–14.
[187] Wang F, Chen W, Zhang C Z, Hua Q, The Kansa method based on the Hausdorff fractal distance for Hausdorff derivative Poisson equations, Fractals 1 (2018), 1–20.
[188] Wang X, Liu X, She K, Zhong S, Pinning impulsive synchronization of complex dynamical networks with various time-varying delay sizes, Nonlinear Anal Hybrid 26 (2017), 307–318.
[189] Wang X, She X, Zhong S, Yang H, Pinning cluster synchronization of delayed complex dynamical networks with nonidentical nodes and impulsive effects, Nonlinear Dyn 88 (2017), 1–12.
[190] Wang Z, Yin K, Jian W, Zhang F, Experimental research on stress relaxation of slip zone soils for Anlesi landslide in Wanzhou city, Chin J Rock Mech Eng 27 (2008) 931–937.
[191] Weron R, Computationally Intensive Value at Risk Calculations, Springer, Berlin, 2004.
[192] Wilkinson S R, Bharucha C F, Fischer M C, et al., Experimental evidence for non-exponential decay in quantum tunnelling, Nature 387 (1997), 575–577.
[193] Xiao R, Yakacki C M, Guo J, et al., A predictive parameter for the shape memory behavior of thermoplastic polymers, J Polym Sci, Polym Phys 54 (2016), 1405–1414.
[194] Xu W, Chen W, Liang Y, et al., A spatial structural derivative model for ultraslow diffusion, Therm Sci 21 (2017), 121–127.
[195] Xu W, Chen W, Liang Y. Feasibility study on the least square method for fitting non-Gaussian noise data, Physica A 2017, 492.
[196] Xu Z B, Decker M, Improving the numerical solution of soil moisture-based Richards equation for land models with a deep or shallow water table, J Hydrometeorol 10 (2010), 308–319.
[197] Yang T, Viscoelastic Mechanics, Huazhong University of Science and Technology Press, Wuhan, 1990.
[198] Yang X, Liang Y, Chen W, A local structural derivative PDE model for ultraslow creep, Comput Math Appl 76 (2018), 1713–1718.
[199] Yanni G, Multi-scale investigation of tensile creep of ultra-high performance concrete for bridge applications, Dissertations & Theses-Gradworks, 2009.
[200] Ye A Q, Gatto R, Colon-Perez L, Mareci T, Morfini G, Magin R L, Using continuous time random walk diffusion to quantify the progression of Huntington's disease, ISMRM 23rd A Meet Exhi, Toronto, Canada, p. 2909, 2015.
[201] Yin D, Zhang W, Cheng C, et al., Fractional time-dependent Bingham model for muddy clay, J Non-Newton Fluid 187 (2013), 32–35.
[202] Yin Q, Wang S, The flabby model of saturated soil establishment and verification, J Heilongjiang Hydrauli Eng Coll 33 (2006), 24–26.
[203] Zähle M, Fractional differentiation in the self-affine case. V-the local degree of differentiability, Math Nachr 185 (2010), 279–306.
[204] Zhao Y, Zhuang X, Ting S J, Gaussian mixture density modeling of non-Gaussian source for autoregressive process, IEEE Trans Signal Process 43 (1995), 894–903.
[205] Zhou Y, Basic Theory of Fractional Differential Equations, World Scientific, Singapore 2014.

Index

Fractional Calculus in Applied Sciences and Engineering

Volume 5
JinRong Wang, Michal Fečkan
Fractional Hermite-Hadamard Inequalities, 2018
ISBN 978-3-11-052220-4, e-ISBN (PDF) 978-3-11-052362-1,
e-ISBN (EPUB) 978-3-11-052244-0

Volume 4
Kecai Cao, YangQuan Chen
Fractional Order Crowd Dynamics. Cyber-Human System Modeling and Control, 2018
ISBN 978-3-11-047281-3, e-ISBN (PDF) 978-3-11-047398-8,
e-ISBN (EPUB) 978-3-11-047283-7

Volume 3
Michal Fečkan, JinRong Wang, Michal Pospíšil
Fractional-Order Equations and Inclusions, 2017
ISBN 978-3-11-052138-2, e-ISBN (PDF) 978-3-11-052207-5,
e-ISBN (EPUB) 978-3-11-052155-9

Volume 2
Bruce J. West
Nature's Patterns and the Fractional Calculus, 2017
ISBN 978-3-11-053411-5, e-ISBN (PDF) 978-3-11-053513-6,
e-ISBN (EPUB) 978-3-11-053427-6

Volume 1
Dingyü Xue
Fractional-Order Control Systems.
Fundamentals and Numerical Implementations, 2017
ISBN 978-3-11-049999-5, e-ISBN (PDF) 978-3-11-049797-7,
e-ISBN (EPUB) 978-3-11-049719-9

Bei Fragen zur Produktsicherheit wenden Sie sich bitte an:
If you have any questions regarding product safety,
please contact:

Walter de Gruyter GmbH
Genthiner Straße 13
10785 Berlin
productsafety@degruyterbrill.com